POCKET NATURE

# V
# ANI

POCKET NATURE

# WILD ANIMALS

CHRIS GIBSON

DK

DORLING KINDERSLEY

LONDON, NEW YORK, MUNICH,
MELBOURNE, AND DELHI

DK LONDON
**Senior Art Editor** Ina Stradins
**Senior Editor** Angeles Gavira Guerrero
**Editor** Cathy Meeus
**Designer** Kenny Grant
**DTP Designer** John Goldsmid
**Production Controller** Linda Dare

**Managing Art Editor** Phil Ormerod
**Managing Editor** Liz Wheeler
**Art Director** Bryn Walls
**Publishing Director** Jonathan Metcalf

DK DELHI
**Designers** Malavika Talukder,
Kavita Dutta
**DTP Designer** Sunil Sharma
**Editors** Dipali Singh, Glenda Fernandes,
Rohan Sinha
**Manager** Aparna Sharma

This edition published in 2010
First published in Great Britain in 2005 by
Dorling Kindersley Limited
80 Strand, London WC2R 0RL

A Penguin Company


A CIP catalogue record for this book
is available from the British Library

ISBN 978 1 4053 4999 4

Reproduced by Colourscan, Singapore
Printed and bound by Sheck Wah Tong, China

see our complete catalogue at
www.dk.com

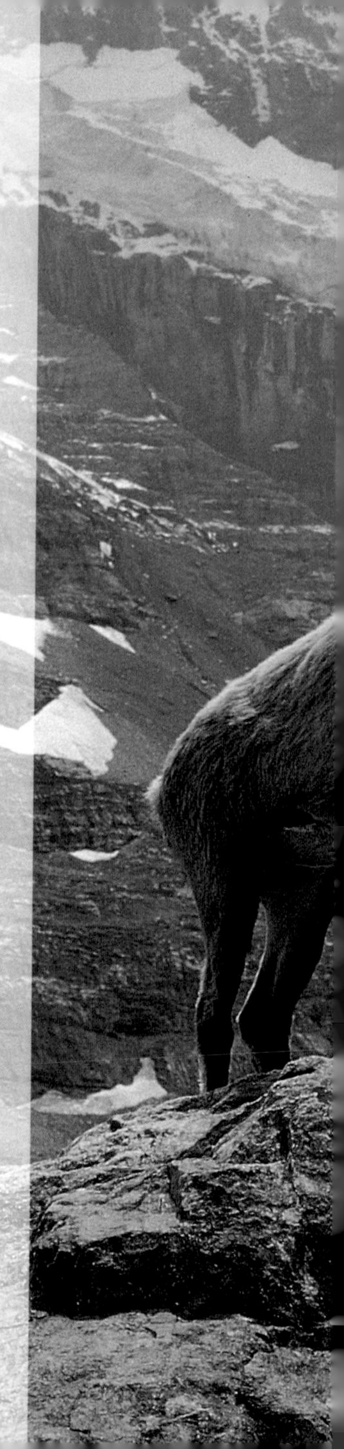

# CONTENTS

a million voices for nature

**The Royal Society for the Protection of Birds** (RSPB) speaks out for birds and wildlife, tackling the problems that threaten our environment. It works with bird and habitat conservation organizations in a global partnership called BirdLife International. Nature is amazing – help us keep it that way.

THE AUTHOR

**Chris Gibson** is a lifelong naturalist who writes, teaches, and broadcasts about the natural world. He is a Senior Conservation Officer for Natural England.

# How this book works

This guide covers in detail 299 of the mammals, reptiles, and amphibians to be found in Europe, with a further 98 referred to briefly. Taken together, these make up 92 per cent of all the species in the region, from the High Arctic to the Mediterranean, and European Russia to the Canary Islands. After a short introduction to the processess of identification, the book is divided into three chapters, one for each group: mammals, reptiles, and amphibians. Within each group, the animals have been grouped by orders, with similar species found together for ease of comparison.

**HABITAT PICTURE**
*Shows the type of habitat in which you are most likely to find the species.*

88

Arc

*Alopex*

Small a usually

FOUND *in Arctic tundra and on sea-ice, in mountains further south, and open woodland in winter.*

front print to 4cm long

hind print smaller

tapered droppings 4–8cm long

SIZE *Body 50–75cm; tail 25–40*
YOUNG *Single litter of up to 18*
DIET *Lemmings, voles, birds and eggs, shellfish, and carrion.*
STATUS *Locally common; rare where over-hunted for its fur.*
SIMILAR SPECIES *None.*

**HABITAT CAPTION**
*Describes the habitat or range of habitats in which the animal is most likely to be found.*

**MAP**
*The shading on the map indicates the potential occurrence of the species in the region, although the species may be occasional in some areas of the range and more prolific in others.*

**SYMBOLS**
*Symbols denote female and male (where visually distinct), and venomous species.*

♀ female

♂ male

venomous

## ▽ GROUP INTRODUCTIONS

Each of the three chapters opens with an introductory page describing the parameters of the group within the European context.

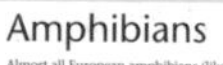

Amphibians

Almost all European amphibians (like the Common Tree-frog pictured below) are dependent on fresh water, at least for breeding, and are usually found in moist habitats. The European total of about 75 amphibian species compares with a world total of about 5,000, a number that is rising as new species are discovered, especially in moist tropical areas. This book covers 53 European species. With the advent of novel techniques of genetic analysis, the relationship between many species and forms is being reconsidered. Thus, new species are continuously being recognized and the list continues to grow.

**PICTURES**
*Photographs of representative specimens show the diversity within the group.*

OCCURS *in extensive deciduous woodland, especially near lakes and rivers.*

Rac

*Nycter*

A small and no It has face, w is accer each ey 1930, t

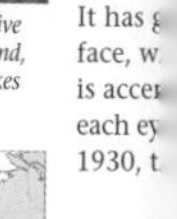

178 REPTILES

Adder

*Vipera berus* (Squamata)

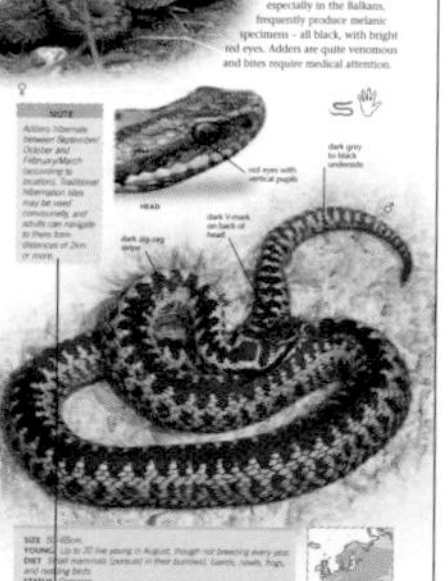

## ▷ FULL-PAGE ENTRIES

Animals that exhibit a more varied or complex range, are of special interest, or are particularly important, are all given full-page entries.

**NOTES**
*Describe unique features, or provide interesting historical or contextual background.*

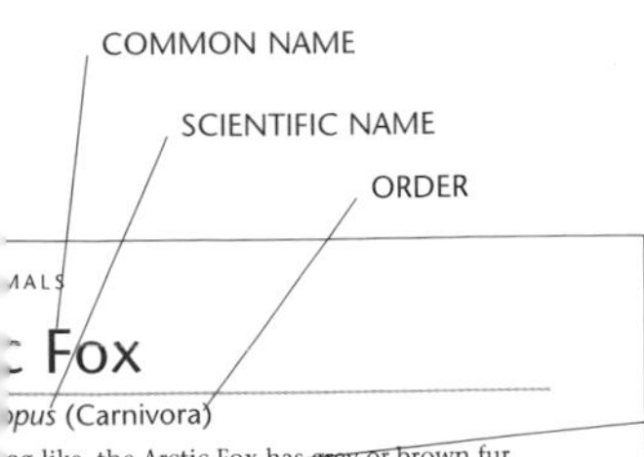


og-like, the Arctic Fox has grey or brown fur, ing white in the winter, which distinguishes it from the larger, redder Red Fox. One form known as the Blue Fox, which is predominant in Iceland but rare elsewhere, has a blue-grey winter coat. By way of adaptation to its cold northern habitats, its ears and muzzle are short and the coat is very thick, both serving to reduce heat loss.

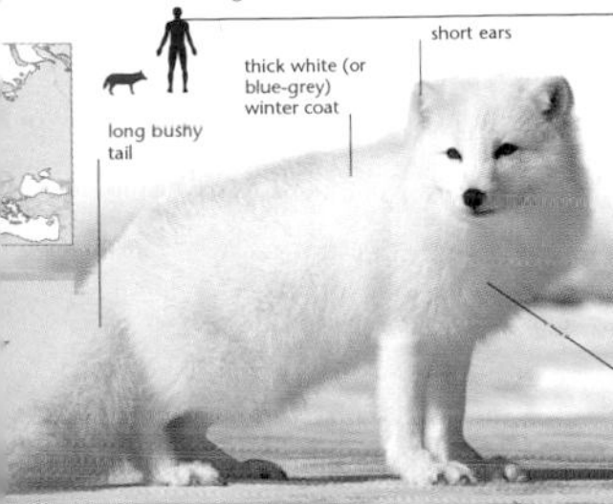


oon Dog

s procyonoides (Carnivora)

cky, long-haired dog, the Raccoon Dog is solitary ial, and (uniquely in the dog family) hibernates. brown fur, darker on the legs and on the It forms a distinctive black mask. The mask ted by a pale grey stripe running from above behind the ears. Introduced from Asia around uropean population has expanded rapidly.

SIZE Body 55–80cm; tail 15–25cm.
YOUNG Single litter of 5–8 or more.
DIET Rodents, worms, amphibians, carrion, tubers, bulbs, fruit, and nuts; scavenges.
STATUS Locally common.
SIMILAR SPECIES Raccoon (p.90); Badger (p.97).

## ▽ SPECIES ENTRIES

The typical page features two entries. Each has a main image of the species, which is taken in the species' natural setting in the wild and is supported by one or more secondary pictures. Annotations, scale artworks, a distribution map, and a data box add key information for each entry.

### DESCRIPTION

*Conveys the main features and the distinguishing characteristics of the animal.*

### SCALE MEASUREMENTS

*Given the vast difference in sizes, two scale drawings, one of a human hand, one of an adult man, are used to give a rough indication of the size of the featured species.*
*The hand represents an average adult hand 18cm long.*

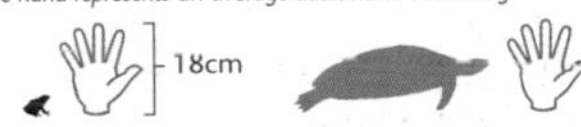


*The man represents an average adult man 1.8m (6ft) tall.*

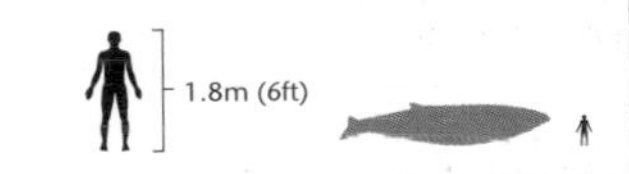


### PHOTOGRAPHS

*Illustrate the species in its natural setting and characteristic pose. Secondary images show different views, sexes, colour variations, or subspecies.*

### COLOUR BANDS

*Bands are colour-coded, with a different colour for each of the three chapters.*

### ANNOTATION

*Characteristic features of the animal are picked out in the annotation.*

### TRACKS AND SIGNS

*These tinted boxes show different aspects of the animal, including footprints, droppings, nests, burrows, and feeding signs. Measurements given are for length unless otherwise stated.*

### OTHER KEY INFORMATION

*These panels provide consistent information on the following points:*
**SIZE:** *body length (including the head); tail length (where present and distinct from the body).*
**YOUNG:** *typical number of eggs or young produced, and the normal breeding season.*
**DIET:** *major components of the diet are listed.*
**ECHOLOCATION:** *for bats only: the typical frequency at which sounds may be heard with a bat detector.*
**STATUS:** *gives the World Conservation Union (IUCN) global conservation status where one has been assigned: Critically Endangered, Endangered, Vulnerable, and Near-threatened (see glossary for definition of terms, and visit www.redlist.org for more information); also gives a more subjective population status relating to the frequency of occurence of the species within its European range: common, locally common (see glossary), scarce, and rare.*
**SIMILAR SPECIES:** *lists species that look similar to the featured animal, often describing distinguishing features.*

# Identifying Mammals

Mammals are the only animals that are covered in fur, although only sparsely so in some groups, especially aquatic ones. They are "warm-blooded" (endothermic), and have the ability to maintain a constant internal body temperature, regardless of changing external conditions. All but a few primitive species, which do not occur in Europe, bear live young rather than lay eggs.

female Rabbit suckling its young

**YOUNG MAMMALS**

*A young mammal is nourished with milk secreted by the mammary glands (modified sweat glands) of its mother. Parental care is one of the key features that sets mammals apart from other animals. Some mammals look after their young until they are several years old.*

## Classification

Mammals are grouped into Orders. There are between 25 and 50 Orders worldwide, according to the opinions of different biologists. Representatives of only 10 are found in the wild in Europe, although one other, the Perissodactyla (odd-toed, hoofed mammals, such as horses), can be found in a semi-wild state. Brief characteristics of each Order are given below.

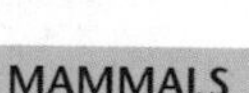

### MAMMALS — CLASS MAMMALIA

**INSECTIVORA** pp.15–28
Feeding on insects, worms, and other invertebrates, these small to medium-sized mammals, have elongate snouts and sharply pointed teeth.

**CHIROPTERA** pp.29–48
This order comprises bats, the only mammals capable of true flight. They are mostly active at night, feeding on insects, located by means of ultrasonic sonar.

**LAGOMORPHA** pp.50–52
Rabbits and hares – small to medium-sized mammals that resemble rodents in many ways. They have well-developed ears and hind legs, and teeth adapted for grazing.

**RODENTIA** pp.53–84
Rodents have teeth adapted to a largely herbivorous lifestyle – a pair of chiselling incisors, a row of grinding cheek teeth, and jaws specialized for gnawing.

**PRIMATES** p.49
With their large brains, keen eyesight, and manual dexterity primates – monkeys and apes – are able to exploit an extremely wide range of food sources.

**CARNIVORA** pp.85–100
In external appearance, carnivores are one of the more diverse orders, linked by meat-eating (often predatory) habits, and powerful teeth and jaws to match.

**PINNIPEDIA** pp.101–104
Although breeding on land, pinnipeds (seals and walruses) are largely aquatic, with a stream-lined shape and feet modified into flippers.

**CETACEA** pp.104–109
With streamlined, hairless bodies, and flukes, flippers, and fins, whales and dolphins are highly adapted to their marine environment.

**ARTIODACTYLA** pp.110–121
Although disparate in size and shape, members of this group always have cloven hooves, and often have paired antlers or horns, especially the males.

**MARSUPIALIA** p.49
A primitive group, native to the southern hemisphere, whose most distinctive feature is that they have pouches in which they nurture their young.

## Habitat and Distribution

Being warm-blooded, mammals can adapt more readily to different climatic conditions than either reptiles or amphibians, and so are found in a wide range of habitats. Knowing which species to expect to see in a specific area is the first step to identification. Reference to the habitat features and distribution map can at least rule out potentially confusing species. However, many animals, especially those that can swim or fly can wander, often over long distances, and so may be found outside their normal range or habitat.

**HABITATS**
*Mammals are found in all European habitats, from Arctic ice to the open sea, the highest mountains to hot, dry grasslands. Many are restricted to just one of these diverse habitat types.*

## Variations

Not all individuals in a species look the same. There may be differences between the sexes; young mammals may be very different from their parents; the appearance may change with the seasons. Furthermore, there may also be variations between animals from different parts of the geographical range. Some of the most common variations are shown, but space does not allow all possible forms to be shown in this book.

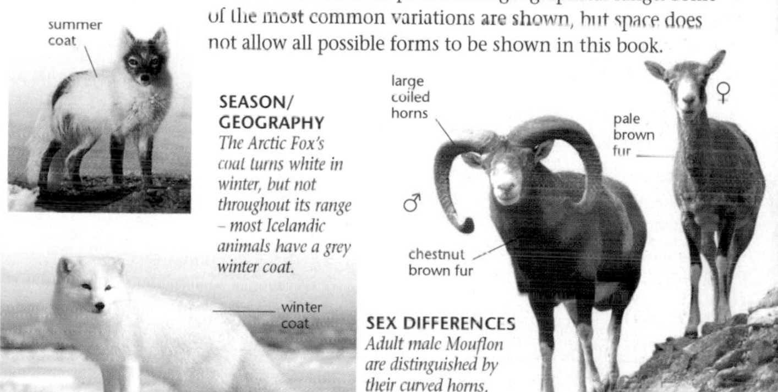


**SEASON/ GEOGRAPHY**
*The Arctic Fox's coat turns white in winter, but not throughout its range – most Icelandic animals have a grey winter coat.*

**SEX DIFFERENCES**
*Adult male Mouflon are distinguished by their curved horns, larger stature, and more patterned coat.*

## Tracks and Signs

Given the often secretive nature, and sometimes nocturnal habits, of many mammals, the best way of recognizing some species is from the signs they leave. These may include droppings, footprints, feeding signs, or distinctive places of shelter, protection, or breeding. Although not always providing conclusive proof of identity, such signs are an important part of the array of identification features.

**TRACKS**
*Tracks may be left in soft mud or snow. They can be difficult to decipher; sometimes only the claws and pads can be seen, but sometimes the entire print, shown as lighter shading, may be present.*

**DROPPINGS**
*The consistency and size of droppings is often characteristic, although variable.*

**FEEDING SIGNS**
*Food remains, even food eaten by several species, may hold important clues.*

# Identifying Reptiles

Reptiles were the first group of animals to evolve a wholly terrestrial way of life. Of the 8,000 or so species living today, few are dependent on wet habitats. All reptiles are "cold-blooded" (ectothermic). Unable to generate heat internally, they obtain sufficient heat from their environment, allowing their bodily functions to operate normally. A consequence of cold-bloodedness is a low feeding requirement: many species are able to fast for extended periods of time.

**REPRODUCTION**
*Most reptiles produce soft-shelled eggs, which are incubated by warmth from the sun or rotting vegetation. Some, however, retain their eggs internally, producing live young.*

**SKIN**
*The texture of the scales that cover the surface of a reptile can impart a characteristic appearance. Detailed examination of the scales, however, is possible only on captive, torpid, or dead specimens.*

GRANULAR

SMOOTH

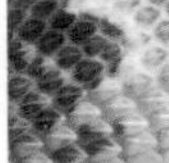

KEELED

HORIZONTAL

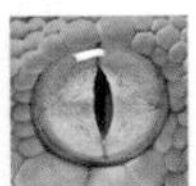

VERTICAL

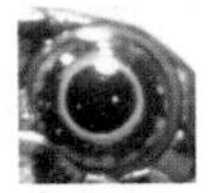

ROUND

**EYES**
*The colour of the eyes and the shape and orientation (pictured left) of the pupil are important identification features if you can get close enough. Furthermore, the colour of the iris is often highly distinctive.*

## Classification

Reptiles are divided into a number of Orders, of which only Chelonia, Amphisbaenia, and Squamata are represented in the wild in Europe. The largest Order, Squamata, is further divided into the limbless Ophidia (snakes) and the Sauria (lizards), most of which possess legs.

### REPTILES — CLASS REPTILIA

**CHELONIA** pp.123–128
Reptiles that have a horny or leathery shell belong to this Order. They include terrestrial, freshwater, and marine species (tortoises, terrapins, and sea turtles).

**AMPHISBAENIA** p.162
Amphisbaenians are a small Order of reptiles, comprising worm-like species that are almost or totally legless; they spend most of their time below ground, and thus are seen only infrequently.

**SQUAMATA** pp.129–181
This Order represents the majority of reptiles, including all the lizards and snakes. Most of these reptiles are predatory and largely diurnal, feeding upon insects, other invertebrates, or small vertebrates. Some of the snakes produce venom which is used to either kill or subdue their prey.

## Habitat and Distribution

Warm, dry habitats tend to be favoured by many reptiles, and as a result the number of species and their abundance increases the further south one goes in Europe. As they have relatively poor powers of dispersal, many distinctive local races, and even species, have developed, often restricted to very small areas.

MEDITERRANEAN SCRUB

Dalmatian Wall Lizard

## Variations

There are several forms of variation possible within a reptile species. Adaptive and geographic variations; sexual differences, in structure and pattern; age differences, the young often strikingly different in colour and pattern; and seasonal differences, with males assuming more vivid colours around the courtship period. There are also black or melanic forms among some species.

Mediterranean Chameleon

**ADAPTIVE VARIATION**
*Some species, most notably the Chameleon, adapt their skin colour and pattern to match their environment, thus providing camouflage.*

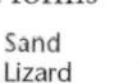

Sand Lizard

**GEOGRAPHIC VARIATION**
*Some species, such as the Sand Lizard, vary greatly in colour and pattern. Every population is variable, and the most frequent form differs from place to place.*

## Behaviour

Reptiles bask to optimize their thermal balance, and this is a good time to observe them. If it is cold they slow down, becoming torpid, which may lead to full hibernation. Unless stated otherwise, species in this book hibernate and are diurnal.

**DEFENCE**
*Defence may include behaviour like playing dead, biting, hissing, and tail shedding, as well as camouflage.*

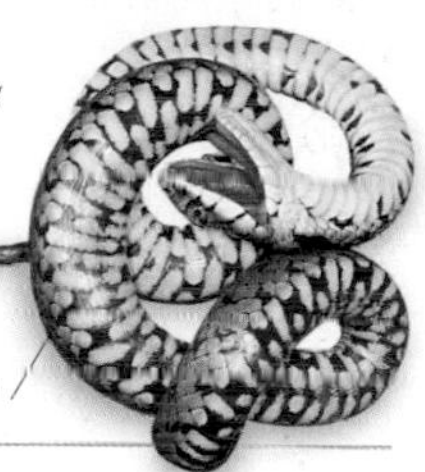

Grass Snake playing dead

## Tracks and Signs

The similarity of different species' tracks means that they are of little value in identification. Likewise, other signs such as hatched eggs, while interesting, are not readily distinguishable. However, the shed skins of lizards and snakes are of great value when attempting to identify a species, retaining characteristic scale textures and patterns, and even traces of colour.

**TRACKS OF A LIZARD**
*Often seen in sandy areas, but rarely attributable to a particular species, the tracks of a lizard include both footprints and tail.*

# Identifying Amphibians

Amphibians were the earliest animal group to exploit terrestrial habitats, but most still lay eggs in water, and the larvae possess external gills for breathing underwater. Being cold-blooded (ectothermic), they become inactive, and often hibernate, when it is cold. Conversely, in hot, dry conditions, they may also reduce their activity, a phenomenon called aestivation.

**SKIN**
*Although toads, such as this Green Toad, tend to have dry, warty skin, most amphibians have smooth, moist skin and are thus found primarily in damp areas.*

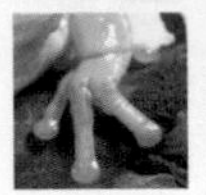

STICKY

WEBBED

**FEET**
*The feet of amphibians show several adaptations to particular lifestyles, sometimes with suction pads for climbing or webbing to aid swimming.*

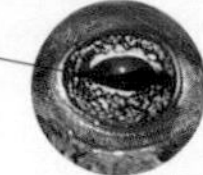


HORIZONTAL

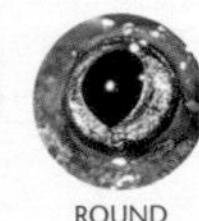

ROUND

VERTICAL

**EYES**
*The form of the eyes, particularly the shape and orientation of the pupil, is a differentiating feature of many groups.*

SMOOTH

WARTY

**SKIN**
*Usually smooth, lacking scales and with a mucus layer, some species may have a characteristic granular covering, which may vary seasonally.*

**TAIL**
*Where present, the tail is used for swimming. In males, the tail may bear a distinctive crest in the breeding season.*

## Classification

There are just three Orders of amphibians, of which one, Caecilia, is exclusively tropical. The remaining two are distinguished by the presence or absence of a tail in the adult. As larvae, the tadpoles of Anura are different from the adults, whereas larval Caudata are more like small adults, though with extended gills.

### AMPHIBIANS — CLASS AMPHIBIA

**ANURA** pp.200–216
The tailless amphibians (frogs and toads) lose their tail in the transition (metamorphosis) from the larval to the adult stages. Many have strongly developed hind legs for jumping and swimming.

**CAUDATA** pp.183–199
Newts and salamanders retain a tail as adults. It is often laterally flattened for more efficient propulsion through water. The legs are generally short and equal in size. Most lay relatively small numbers of eggs, usually attaching them to underwater leaves or rocks.

## Habitat and Distribution

Being cold-blooded and water-dependent, the greatest diversity of amphibians is found in the southern part of Europe, but away from the drier habitats. As a result of the often lumbering gait of many amphibians, and their sometimes very precise habitat requirements, the world range of some species is extremely small, with consequent risks of extinction.

## Variations

The appearance of many species is often variable in colour, pattern, and structure. Such differences may be geographical, sexual, seasonal, or age-related, and can pose challenges to identification. Where such differences are apparent, it is invariably the adult males in breeding condition which have the brightest colours, and most flamboyant crests.

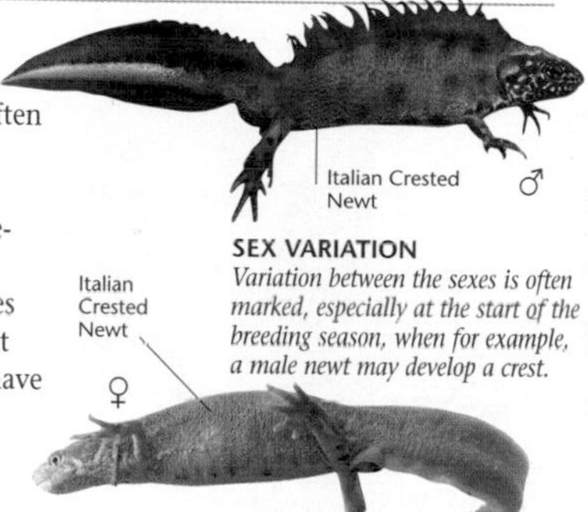


**SEX VARIATION**
*Variation between the sexes is often marked, especially at the start of the breeding season, when for example, a male newt may develop a crest.*

## Behaviour

Amphibians are active by day and night, especially when breeding. Many newts have complex mating rituals, and the mating frenzies of some frogs and toads can be dramatic. Defensive behaviour includes threatening postures, display of warning colours, and secretion of toxins from the skin.

Common Spadefoot

**REFUGE**
*In unfavourably cold or dry conditions, many amphibians become inactive and take refuge in moist microhabitats, under stones and logs, or buried in mud or sand.*

## Tracks and Signs

Amphibians leave rather little in the way of diagnostic tracks and signs. Apart from hearing the calls or finding the early developmental stages, the only realistic way of recording their presence is to locate the adults. This may require considerable time (some are very secretive) and patience (others dive for refuge as soon as they detect a threat and can remain submerged for hours).

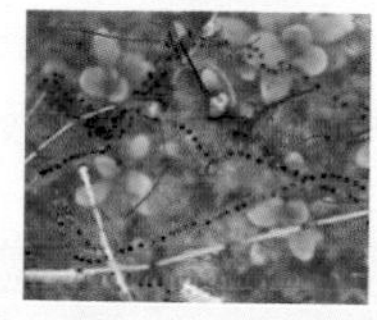

**TOAD SPAWN**
*The eggs (spawn) and larvae (tadpoles), of amphibians are distinctive within a group but cannot easily be identified to the species level.*

**VOCALIZATION**
*Many frogs and toads produce complex and distinctive songs, especially in the breeding season. They are difficult to describe, and it is best to obtain a sound recording guide.*

# Mammals

There are about 4,675 species of mammals worldwide, ranging from the Pygmy White-toothed Shrew (p.25), a little over 3cm long, to the colossal 30m-long Blue Whale. Around 230 species (including the Fallow Deer, pictured below) are found in Europe, of which 157 are covered in this book. Cetaceans – whales, dolphins, and porpoises – are infrequently seen from land, and so only the most commonly sighted ones are included.

The species are first grouped into orders, and then by genus, keeping species with similarities in appearance or behaviour together, for ease of comparison.

STOAT

WHITE-BEAKED DOLPHIN

LONG-EARED BAT

BROWN HARE

# Western Hedgehog

*Erinaceus europaeus* (Insectivora)

**FOUND** *in lowland grassland and open woods; very frequently in gardens where they perform a valuable role in pest control.*

Hedgehogs, characterized by their covering of protective spines, up to 3cm long, and short dark legs, are familiar animals over much of Europe, having colonized even the largest cities. The three hedgehog species barely overlap in distribution, with the Western Hedgehog occupying western and northern Europe. The Western Hedgehog's underside is clothed in uniformly-coloured fur: dark brown in those that inhabit the north; pale cream in those in the south. A noisy forager and largely nocturnal, this hedgehog hibernates in nests of dry grass and leaves.

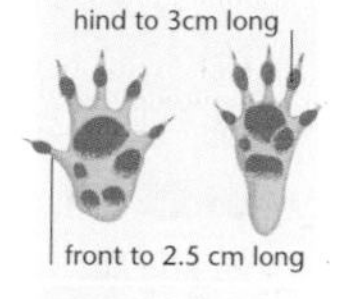


droppings to 4cm long

**NOTE**

*Spines are an effective defence against most predators – apart from motor vehicles. Sadly, road casualties are all too common.*

face and limbs hidden

**ROLLED DEFENSIVE POSTURE**

covered in spines above

narrow spine-free "parting"

creamy brown spines often dark-tipped

**SIZE** *Body 20–30cm; tail 1–4cm.*
**YOUNG** *One or two litters of up to six; June–September.*
**DIET** *Earthworms, slugs, beetles, and other invertebrates; birds' eggs and nestlings; carrion; some plant material and fungi.*
**STATUS** *Common.*
**SIMILAR SPECIES** *Algerian Hedgehog (p.16), which is smaller and paler; Eastern Hedgehog (p.16), which lacks a uniformly-coloured underside.*

# Eastern Hedgehog

*Erinaceus concolor* (Insectivora)

**INHABITS** *most lowland habitats, especially grassland and woodland up to 1,400m, and also urban areas.*

The eastern counterpart of the very similar Western Hedgehog, the Eastern Hedgehog differs in its distribution and the contrasting colour of its underside fur, the throat being almost white. Like other hedgehogs, its tail is barely visible. It is largely nocturnal, and hibernates, at least in the northern part of its range.

**SIZE** *Body 20–30cm; tail up to 3cm.*
**YOUNG** *One or two litters of up to six.*
**DIET** *Earthworms, slugs, beetles, and other invertebrates; carrion.*
**STATUS** *Locally common.*
**SIMILAR SPECIES** *Western (p.15) and Algerian Hedgehogs (below).*

# Algerian Hedgehog

*Atelerix algirus* (Insectivora)

**FORAGES** *in lowland grassland, cultivated areas, as well as in gardens; favours dry Mediterranean scrubland.*

Possibly introduced from North Africa, the Algerian Hedgehog is smaller than other European hedgehogs. Its limited range overlaps with that of the darker Western Hedgehog in Spain, although an examination of the crown parting may be necessary to confirm identification. Unlike its relatives, this hedgehog does not hibernate. It is largely nocturnal.

**SIZE** *Body 20–25cm; tail 2–4cm.*
**YOUNG** *Poorly known.*
**DIET** *Ground-dwelling invertebrates and small vertebrates, carrion, and fungi.*
**STATUS** *Rare; locally common.*
**SIMILAR SPECIES** *Western (p.15) and Eastern (above) Hedgehogs.*

# Common Shrew

*Sorex araneus* (Insectivora)

Like all shrews, the Common Shrew is a small, active but secretive animal, its elongated snout bearing long sensory bristles. Adults are distinctly three-coloured, with brown back, creamy underside, and chestnut flanks. In contrast, juveniles have a less distinct flank coloration, a paler overall appearance, and a thicker tail, which bears tufts of bristly hair. Common Shrews are active by day and night, throughout the year, a consequence of their very high metabolic rate, which means they must eat more than 90 per cent of their body weight each day.

FREQUENTS *all habitats with significant ground cover, particularly rough grassland.*

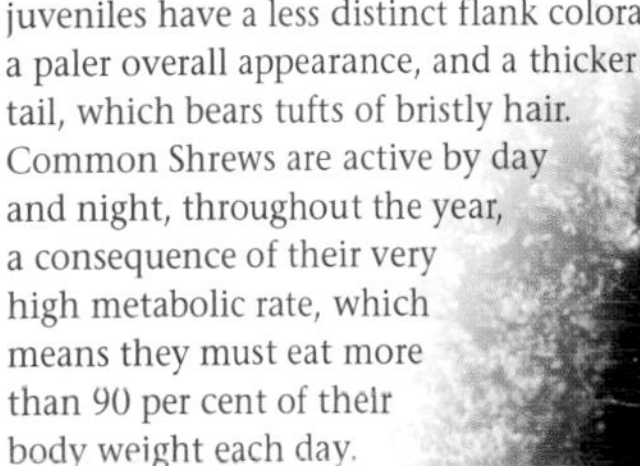


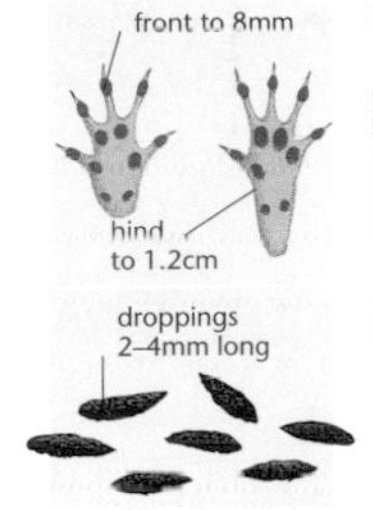


**NOTE**

*Shrews are relatively unpalatable, and so are frequently killed but not eaten by predators such as cats. This provides a good opportunity for close examination of critical identification features.*

dark brown back

small, inconspicuous eyes and ears

reddish flanks

**SIZE** *Body 5.5–9cm; tail 3–6cm.*
**YOUNG** *Up to four litters of 6–7, April–August.*
**DIET** *Worms, slugs, woodlice, spiders, beetles, and other ground-dwelling invertebrates; small amounts of seed.*
**STATUS** *Common.*
**SIMILAR SPECIES** *Millet's (p.18), Spanish* (S. granarius), *and Appennine* (S. samniticus) *Shrews, which have different distributions.*

# Millet's Shrew

*Sorex coronatus* (Insectivora)

**OCCUPIES** *scrub and well-vegetated grass habitats, preferring drier and warmer sites than those frequented by the Common Shrew.*

The western continental counterpart of the Common Shrew, Millet's Shrew is usually a little smaller, but definite identification relies on skull measurements, chromosome counts, and distribution. Millet's Shrew has reddish brown flanks, pale underparts, tail up to two-thirds the length of head and body, and inconspicuous eyes and ears.

**SIZE** *Body 6–8cm; tail 3.5–4.5cm.*
**YOUNG** *Up to four litters of 3–7.*
**DIET** *Ground-dwelling invertebrates; seeds.*
**STATUS** *Common.*
**SIMILAR SPECIES** *Common (p.17) and Spanish* (S. granarius) *Shrews, which are best distinguished by distribution.*

# Masked Shrew

*Sorex caecutiens* (Insectivora)

**OCCURS** *in northern coniferous forest and associated open habitats, such as moorland and tundra.*

Also known as Laxmann's Shrew, the Masked Shrew is a medium-sized shrew of northern forests and tundra. It is a bi-coloured species, with a sharp demarcation between the brown upperparts and whitish underparts, and the feet are covered with silvery white hair.

**SIZE** *Body 4.5–7cm; tail 3–4.5cm.*
**YOUNG** *Up to four litters of 7–8.*
**DIET** *Beetles, other ground invertebrates, and pine seeds.*
**STATUS** *Common.*
**SIMILAR SPECIES** *Pygmy Shrew (p.20), which is paler above.*

# Alpine Shrew

*Sorex alpinus* (Insectivora)

A large, dark grey shrew, the underparts of the Alpine Shrew are only very slightly paler than its upperparts. This lack of contrast, its relatively long tail, and montane habitat make this one of the more easily recognized shrews. Unlike most species, which are largely restricted to the ground, Alpine Shrews are excellent climbers, using their tail for support and balance, although they do also hunt on the ground, particularly around rocky mountain streams.

**INHABITS** *coniferous forest, meadows, and open rocky habitats, to more than 3,000m, in the mountains of central Europe.*

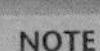

**NOTE**

*The Alpine Shrew has the longest tail, relative to its head and body, of any European shrew. This characteristic is linked to its strongly arboreal habit and, therefore, its need to maintain balance.*

**SIZE** *Body 6–7.5cm; tail 6–7.5cm.*
**YOUNG** *Two or three litters of 5–6.*
**DIET** *Snails, earthworms, spiders, insects, and other invertebrates.*
**STATUS** *Locally common.*
**SIMILAR SPECIES** *Dusky Shrew (p.21), which has a shorter tail and different distribution.*

# Pygmy Shrew

*Sorex minutus* (Insectivora)

**FOUND** *in a very wide range of lowland and upland grassy, heath and scrub habitats; more tolerant of shorter vegetation than the Common Shrew.*

The Pygmy Shrew is distinctly smaller than most other widespread shrews in Britain and Europe. Its upper fur is a relatively pale, medium brown colour, which merges into the whitish underside without a distinct flank colour. The tail is relatively long, greater than two-thirds of the head and body length, and broad, covered as it is by a dense clothing of hairs. Active by both day and night, with alternating periods of foraging and rest, the Pygmy Shrew is rather more diurnal than the Common Shrew, although more often heard squeaking than seen.

droppings to 4mm long

front print to 8mm long

hind print to 1.2cm long

**NOTE**

*Teeth characteristics are useful for identifying shrews. In Europe, the Pygmy Shrew is unique in that its third single-cusped tooth in the upper jaw is longer than the second.*

no distinct flank colour

very small ears, hidden in fur

slender, pointed muzzle

**SIZE** *Body 4–6cm; tail 3–4.6cm.*
**YOUNG** *Two litters of 4–7; April–August.*
**DIET** *Ground-dwelling invertebrates, especially beetles and woodlice.*
**STATUS** *Common, though usually less abundant than Common Shrews in the same habitats.*
**SIMILAR SPECIES** *Common Shrew (p.17); Least Shrew (p.21), which is even smaller and has a shorter tail.*

# Least Shrew

*Sorex minutissimus* (Insectivora)

Even smaller than the Pygmy Shrew (p.20) and with a shorter and narrower tail, the Least Shrew is found in the northern forest zone, stretching eastwards from Scandinavia. Least Shrews show longer periods of activity than their relatives, and need to consume between two and five times their body weight in food each day.

**INHABITS** *open and wet patches, including peat bogs, in and around northern coniferous forests.*

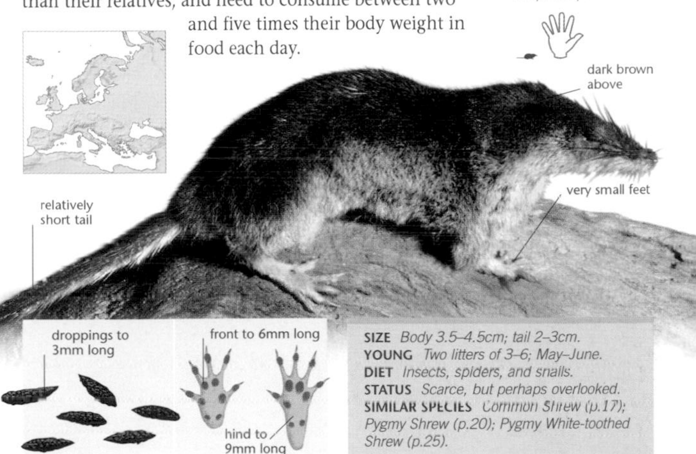


droppings to 3mm long

front to 6mm long

hind to 9mm long

**SIZE** *Body 3.5–4.5cm; tail 2–3cm.*
**YOUNG** *Two litters of 3–6; May–June.*
**DIET** *Insects, spiders, and snails.*
**STATUS** *Scarce, but perhaps overlooked.*
**SIMILAR SPECIES** *Common Shrew (p.17); Pygmy Shrew (p.20); Pygmy White-toothed Shrew (p.25).*

# Dusky Shrew

*Sorex isodon* (Insectivora)

The Dusky Shrew is a large shrew found in parts of the northern forest zone, hence its alternative name of Taiga Shrew. Its scientific name *isodon*, or "equal teeth", refers to the regular decrease in size of its single-cusped teeth, from front to back. It is one of few species, and the only one in its range, which has underparts almost the same colour as the upperparts, and has no distinct flank colour.

**LIVES** *in damp northern coniferous forests and grassland, especially with a thick moss layer.*

**SIZE** *Body 5–8cm; tail 4–5.5cm.*
**YOUNG** *Two or three litters of 6–7; June–August.*
**DIET** *Insects, worms and other invertebrates, and small amphibians.*
**STATUS** *Rare.*
**SIMILAR SPECIES** *Alpine Shrew (p.19).*

droppings to 4mm long

hind to 1.4cm long

front to 1cm long

**FOUND** *in a variety of aquatic habitats, from seaweed-covered boulders to mountain streams up to 2,500m; rarely wanders to more terrestrial sites.*

# Water Shrew

*Neomys fodiens* (Insectivora)

A large species, the Water Shrew is usually found in damp areas, as its name would suggest. It is a good swimmer, aided by fringes of hair on the feet and a keel of silvery hairs on the tail. When seen underwater, this shrew has a silvery appearance due to bubbles of air trapped in its fur. On land, it is black above and usually white below, with a sharp demarcation between the two. However, there is variation in the underside colour, and some individuals are black all over. Water Shrews are largely nocturnal foragers, leaving piles of food remains around their feeding sites. They build tunnels near water and live in nests made of dry grass and leaves.

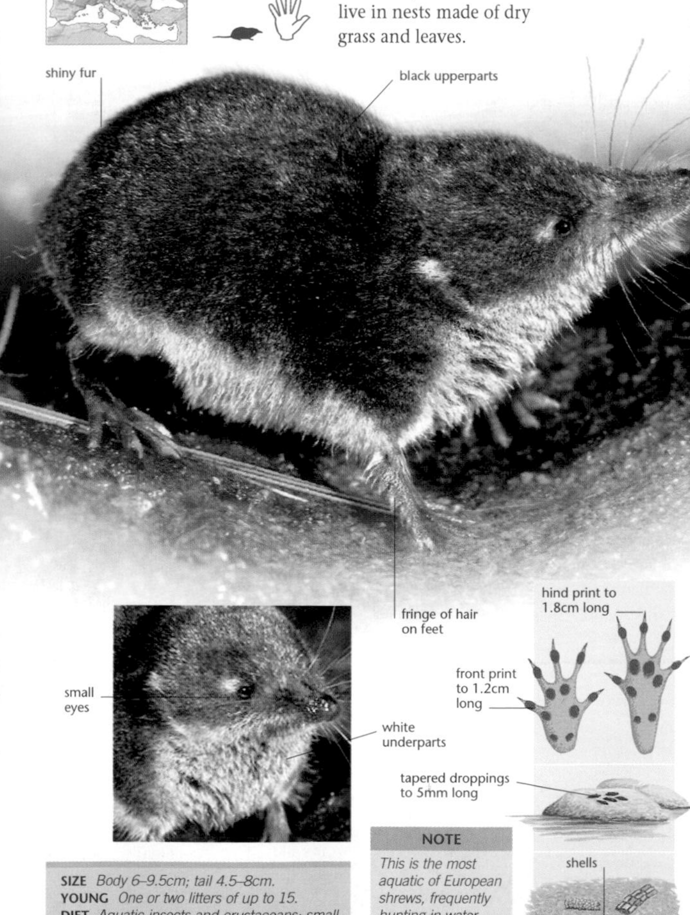


**SIZE** *Body 6–9.5cm; tail 4.5–8cm.*
**YOUNG** *One or two litters of up to 15.*
**DIET** *Aquatic insects and crustaceans; small fish and amphibians.*
**STATUS** *Locally common.*
**SIMILAR SPECIES** *Miller's Water Shrew (p.23); Pyrenean Desman (p.26).*

**NOTE**

*This is the most aquatic of European shrews, frequently hunting in water, where it feeds on fish and frogs after paralysing them with toxic saliva.*

**FOOD REMAINS**

# Miller's Water Shrew

*Neomys anomalus* (Insectivora)

Similar to the Water Shrew, but generally smaller and lacking the complete fringe of hair on its tail, Miller's Water Shrew is scattered across Europe. Although it is adept at swimming, it is less confined to aquatic habitats than its close relative. In western Europe, it is found largely in mountains, but frequently occupies lowland habitats in the east.

black fur above

contrasting white fur beneath

**LIVES** *in still or flowing waters, bogs, and mountain streams; also damp woodland and grassland.*

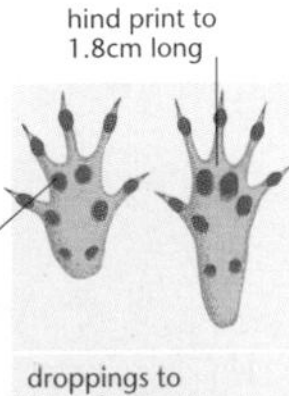


droppings to 5mm long

**SIZE** *Body 6.5–8.5cm; tail 4–6.5cm.*
**YOUNG** *One or two litters of about six.*
**DIET** *Aquatic and terrestrial insects; small fish and amphibians.*
**STATUS** *Locally common.*
**SIMILAR SPECIES** *Water Shrew (p.22); Pyrenean Desman (p.26).*

# Bicoloured White-toothed Shrew

*Crocidura leocodon* (Insectivora)

The Bicoloured White-toothed Shrew has all the features of its group, with a lack of pigmentation on its teeth, prominent ears, and velvety fur. It is, however, the only one with a clear line of demarcation between the grey-brown upperparts and creamy white underparts, which also extends to its relatively short, stout tail

**OCCUPIES** *lowland scrub, woodland edges, gardens, and agricultural land, favouring dry habitats.*

dense grey-brown upper fur

pinkish, pointed muzzle

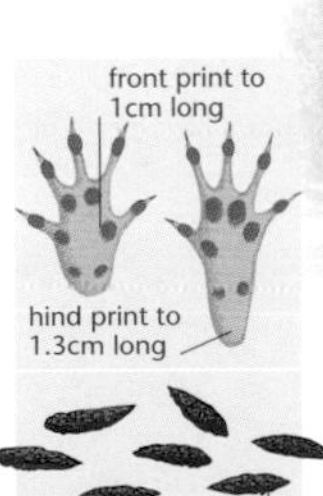


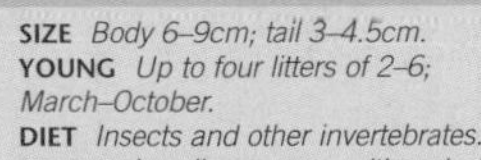

**SIZE** *Body 6–9cm; tail 3–4.5cm.*
**YOUNG** *Up to four litters of 2–6; March–October.*
**DIET** *Insects and other invertebrates.*
**STATUS** *Locally common, although declining.*
**SIMILAR SPECIES** *All white-toothed shrews (pp.24–25).*

# Greater White-toothed Shrew

**FOUND IN** *lowland, dry grassland, woodland, and cultivated land; also gardens, houses, and farm buildings.*

*Crocidura russula* (Insectivora)

The white-toothed shrews are superficially similar to the red-toothed species, but can usually be told apart as a group by their prominent larger ears, not hidden in fur, and their shorter tail, covered with whiskery hair. But differentiation of species is more difficult, especially between Greater and Lesser White-toothed Shrews. Size is some indication, but there is considerable overlap, and details of the teeth may be needed for definite identification. The Greater White-toothed Shrew is the shrew that most frequently enters buildings, especially during the winter months.

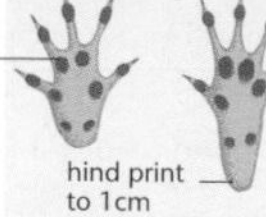


droppings 2–4mm long

**NOTE**

*This species may sometimes be seen "caravanning", when litters of young are led to safety by their mother, each grasping the tail of the animal in front with their teeth.*

mid-brown above

prominent ears

**SIZE** *Body 5.5–8.5cm; tail 2.5–4.5cm.*
**YOUNG** *Four or five litters of 3–4, February–November.*
**DIET** *Insects and other invertebrates; small vertebrates and carrion.*
**STATUS** *Common.*
**SIMILAR SPECIES** *Other White-toothed Shrews, especially Lesser (p.25); related island species are* Crocidura canariensis *(E. Canaries),* C. osorio *(Gran Canaria),* C. sicula *(Sicily, Malta), and* C. zimmermanni *(Crete).*

# Lesser White-toothed Shrew

*Crocidura suaveolens* (Insectivora)

While examining the teeth of a live shrew is difficult, a definitive separation between Lesser and Greater White-toothed Shrews can be made in a dead specimen. The second single-cusped tooth in the upper jaw of the Lesser is smaller than the third, whereas they are of a similar size in the Greater. The Lesser is also less active during daylight hours.

**FREQUENTS** *dry grassland, woodland, and cultivated areas, covered with bracken or other tall vegetation.*

"whiskered" tail

pale underside

reddish brown fur

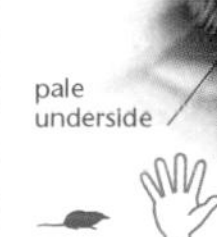

**SIZE** *Body 5.5–7cm; tail 3–4cm.*
**YOUNG** *Up to four litters of about six, February–October.*
**DIET** *Insects and other ground invertebrates.*
**STATUS** *Scarce and declining.*
**SIMILAR SPECIES** *Greater White-toothed Shrew (p.24).*

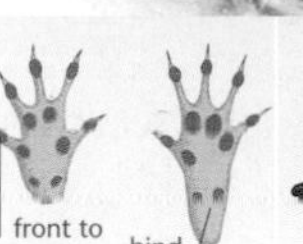


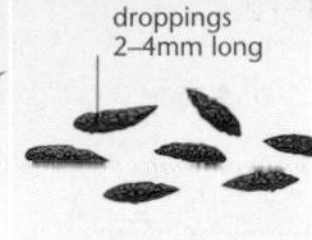


# Pygmy White-toothed Shrew

*Suncus etruscus* (Insectivora)

Despite its tiny size (it is one of the smallest mammals in the world), the Pygmy White-toothed Shrew is a ferocious predator, feeding mainly on large insects such as grasshoppers and crickets that are similar in size to itself. Apart from its size, it can be distinguished from other white-toothed shrews by the presence of four rather than three single-cusped teeth.

**OCCUPIES** *hot, dry lowland habitats, including grassland, scrub, open woodland, and gardens.*

grey-brown above

small hind feet

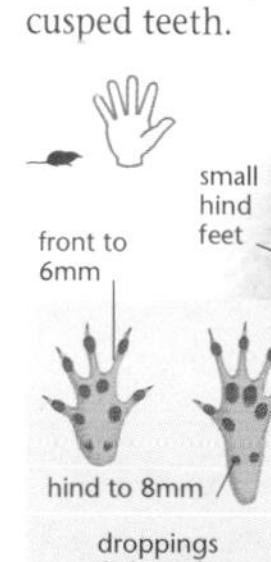

**SIZE** *Body 3.5–5cm; tail 2.5–3cm.*
**YOUNG** *Up to four litters of 2–5; March–October.*
**DIET** *Ground-dwelling invertebrates.*
**STATUS** *Scarce to rare.*
**SIMILAR SPECIES** *Least Shrew (p.21), which is a northern forest species.*

# Pyrenean Desman

*Galemys pyrenaicus* (Insectivora)

**LIVES** *in and around aquatic habitats in northern Iberia, from sea shores to high mountain streams and lakes; look out for small, twisted droppings.*

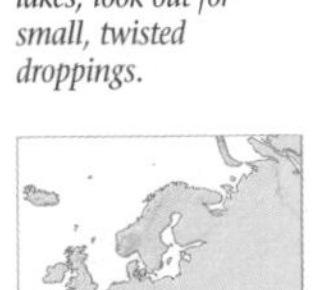

Largely aquatic in habit, the Pyrenean Desman has webbed feet that it uses as paddles and a flattened tail it uses as a rudder. Besides these adaptations to its lifestyle, it also has strong claws for anchorage in fast-flowing streams. Its sensitive muzzle is used to search out invertebrates, but its eyes are small and its vision poor. Desmans are largely nocturnal, and much of their activity is guided by an acute sense of smell. The conspicuous scent gland at the base of the tail produces an odour, which is detectable even by human noses.

large, webbed hind feet

long, flexible, flattened snout

**NOTE**

*Pyrenean Desmans are rarely observed, due to their secretive nature, largely nocturnal habits, and often remote habitats. Perhaps not surprisingly, the species was discovered only as recently as 1811.*

dark brown fur

small eyes

flattened, long tail

paler below

strong claws

**SIZE** *Body 11–15cm; tail 12–15cm.*
**YOUNG** *One or two litters of up to four; February–July.*
**DIET** *Aquatic invertebrates.*
**STATUS** *Vulnerable; scarce and declining.*
**SIMILAR SPECIES** *Water Shrews (pp.22–23), which are considerably smaller.*

front to 1.5cm

hind to 3cm

twisted droppings to 1cm long

# Common Mole

*Talpa europaea* (Insectivora)

Common Moles are rarely seen above ground, but their presence is easily noted from the spoil heaps (molehills) resulting from their burrowing activity, which continues day and night throughout the year. They are least visible in the heat of summer when their invertebrate prey is forced down to deeper, more moist soil levels. The European species of mole are difficult to tell apart, but fortunately the distributions have little overlap, and the Common Mole is the only species to be found over much of the area.

**FAVOURS** *grassland, cultivated areas, and deciduous woodland, up to 2,000m altitude; mainly montane in the south; avoids places that are prone to flooding.*

**NOTE**

*Moles are adapted to burrowing, with a cylindrical body shape, spade-like forefeet, ears that do not project, and poor vision, but a keen sense of smell.*

large front feet

dark velvety fur

small, but open, eyes

hind to 2cm

formed by claws only, each front print is up to 4mm long

spoil heap from tunnelling

**MOLEHILL**

**SIZE** *Body 11–16cm; tail 2–4cm.*
**YOUNG** *Usually a single litter of 3–4; May–June.*
**DIET** *Earthworms, insect larvae, and other subterranean invertebrates, often paralysed with a bite and stored alive for times of shortage.*
**STATUS** *Common.*
**SIMILAR SPECIES** *Blind Moles (p.28); Roman Mole* (T. romana), *both of which have permanently closed eyes.*

**INHABITS** *grassland and deciduous woodland, mostly in the uplands, often in drier areas than other moles.*

# Blind Mole

*Talpa caeca* (Insectivora)

Aptly named from the fact that its eyes are permanently covered with a membrane, the Blind Mole is smaller than other European moles, and especially favours upland habitats, up to 2,000m. It also has distinct tooth characteristics, in particular its central incisors being almost twice the size of the lateral ones. The Blind Mole has a more slender muzzle than its relatives, and a relatively constant feature is the presence of whitish hair on its tail, legs, and lips.

**NOTE**

*Three further "blind" moles are listed below as similar species. All three are very similar and best identified by their distribution – Iberian Peninsula, Italy, and the Balkans respectively.*

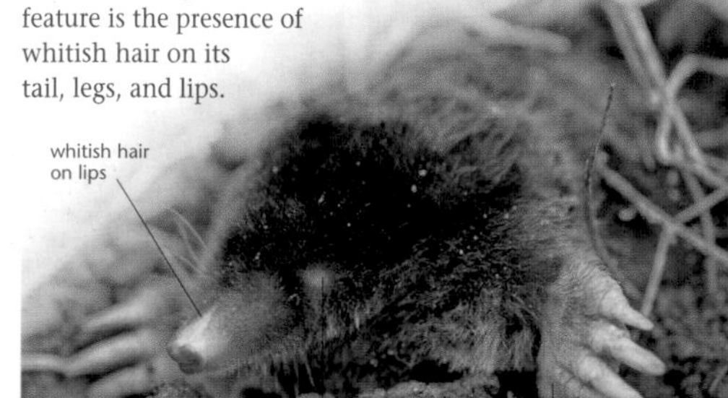


**SIZE** *Body 9.5–14cm; tail 2–3cm.*
**YOUNG** *Single litter of 3–4; May–June.*
**DIET** *Subterranean invertebrates.*
**STATUS** *Locally common.*
**SIMILAR SPECIES** *Common Mole (p.27); Iberian Mole* (T. occidentalis)*; Roman Mole* (T. romana)*; Balkan Mole* (T. stankovici).

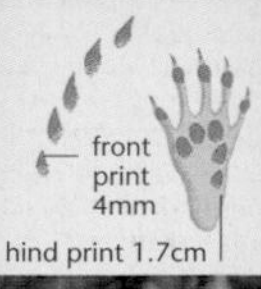

# Lesser Horseshoe-bat

*Rhinolophus hipposideros* (Chiroptera)

Often forming large communal roosts, the Lesser Horseshoe-bat is the smallest species of its group in Europe. Winter roosts, in which it hibernates between November and March, are largely underground, in caves, tunnels, and cellars, but breeding colonies are mostly to be found in buildings. As with other horseshoe-bats, the wings are wrapped around its body while at roost (completely so in the case of the Lesser and Greater Horseshoe-bats). Individuals hang separately from the ceiling, without touching their neighbours.

**FOUND** *in well-wooded areas, usually associated with limestone geology, in warmer regions of Europe; hunts close to the ground, usually below 5m.*

**NOTE**

*The five European horseshoe-bats are characterized by complex folds of skin on the face, which are involved in the production of ultrasound for echolocation; they also differ in details of the sella and lancet.*

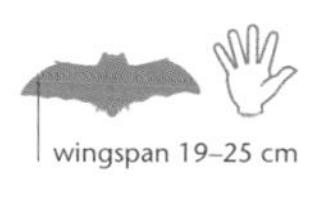


wings enfold the body completely at rest

greyish ears without a tragus

grey-brown above

tapering sella

pale grey below

pale brown membrane

droppings to 8mm long

**SIZE** *Body 3.5–4.5cm; tail 2.5–3.3cm.*
**YOUNG** *Single young; June–August.*
**DIET** *Small insects and spiders.*
**ECHOLOCATION** *110kHz.*
**STATUS** *Vulnerable; scarce and declining.*
**SIMILAR SPECIES** *Mediterranean (p.30) and Blasius' (p.31) Horseshoe-bats, which are paler above.*

# Greater Horseshoe-bat

*Rhinolophus ferrumequinum* (Chiroptera)

**OCCURS** *in wooded areas, near roosting sites and pastures; leaves insect remains below perches in trees and cave entrances.*

The largest European horseshoe-bat, this species has a more butterfly-like flight, interspersed with glides, than the other horseshoe-bats. It forms large breeding colonies in buildings, often with 500 or more females, and roosts underground in winter.

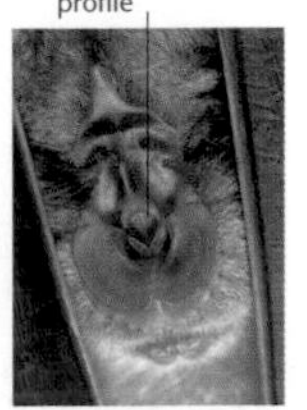


FACE

wingspan 29–35cm

greyish brown above

fluffy fur

**SIZE** *Body 5.5–7cm; tail 3.5–4cm.*
**YOUNG** *Single young; June–August.*
**DIET** *Large flying insects.*
**ECHOLOCATION** *85kHz.*
**STATUS** *Locally common, but declining.*
**SIMILAR SPECIES** *Mehely's Horseshoe-bat (p.31), which is paler and has longer ears.*

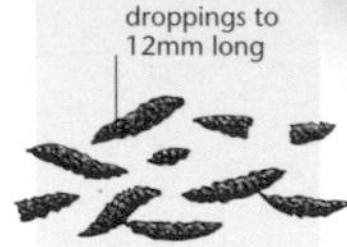

# Mediterranean Horseshoe-bat

*Rhinolophus euryale* (Chiroptera)

**FREQUENTS** *wooded areas, usually in hills and mountains, with access to underground roosting sites.*

Despite its distinctive slow, fluttering flight and frequent hovering, this horseshoe-bat is not easy to identify. It is a widespread southern species, which is less dependent on buildings and human settlements than most other horseshoe-bats. Roosts, both in winter and for breeding, are usually in caves.

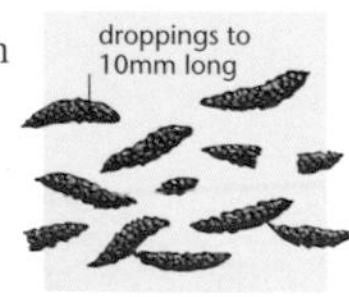


slender sella, pointed in profile

wingspan 30–32cm

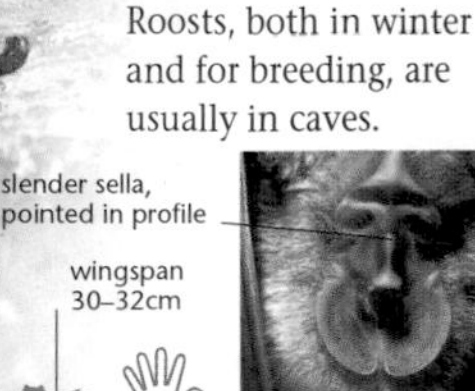

FACE

relatively large ears

reddish or lilac tinged fur above

**SIZE** *Body 4.5–5.8cm; tail 2–3cm.*
**YOUNG** *Single young; June–August.*
**DIET** *Moths and other flying insects.*
**ECHOLOCATION** *102kHz.*
**STATUS** *Vulnerable; locally common, but declining in the north of its range.*
**SIMILAR SPECIES** *Other horseshoe-bats.*

# Blasius' Horseshoe-bat

*Rhinolophus blasii* (Chiroptera)

The least well-known member of its group, Blasius' Horseshoe-bat is intermediate between other species. However, it is much paler underneath than the other species, although this is difficult to see, except at roost, because of its nocturnal foraging habits. Roosts are formed mainly in caves, but towards the north of its range, nursery roosts may be found in buildings.

**INHABITS** *limestone areas, with woodland and scrub, and access to roosting sites in caves or buildings.*

constricted sella

almost white beneath

broad nose-leaf

wingspan 27–31cm

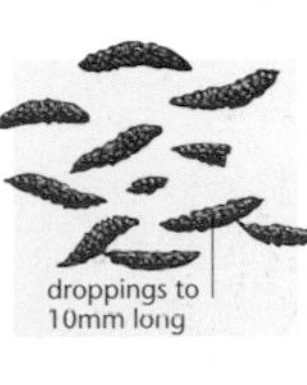


**SIZE** *Body 4.5–5.5cm; tail 2.5–3cm.*
**YOUNG** *Single young; June–August.*
**DIET** *Moths and other insects.*
**ECHOLOCATION** *95kHz.*
**STATUS** *Near-threatened; rare and declining.*
**SIMILAR SPECIES** *Mediterranean Horseshoe-bat (p.30); also other horseshoe-bats.*

# Mehely's Horseshoe-bat

*Rhinolophus mehelyi* (Chiroptera)

A relatively large and pale species, Mehely's Horseshoe-bat has a slow, gliding flight. There is a sharp demarcation between its grey-brown upper- and paler underparts. In its cave roosts, the wings do not completely envelop the body. It is a poorly known species, distributed in a number of discrete populations through its range, many of which are declining.

**FOUND** *mainly in limestone areas, especially with access to water, and large caves for roosting.*

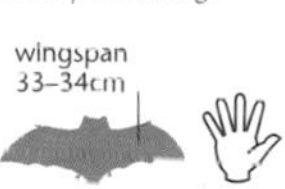

**ROOST SITE**

**SIZE** *Body 5.5–6.5cm; tail 2.5–3cm.*
**YOUNG** *Single young; June–August.*
**DIET** *Moths and other insects.*
**ECHOLOCATION** *105kHz.*
**STATUS** *Vulnerable, rare, and declining.*
**SIMILAR SPECIES** *Greater Horseshoe-bat (p.30), which has shorter ears.*

# Daubenton's Bat

*Myotis daubentonii* (Chiroptera)

**INHABITS** *open wooded areas, with access to water for feeding; summer roosts in cracks in trees and buildings, and under bridges.*

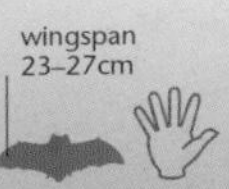


Small to medium-sized, Daubenton's Bat often feeds low over still or slow-moving water. It swims well and is capable of taking flight off the surface of water. In winter, it roosts in underground sites with high humidity, where torpid specimens often covered in dew are found. It occupies rock crevices and may also be found among screes on cave floors. One of the most numerous European bats, it is showing some signs of increase in population, perhaps as a result of climatic change.

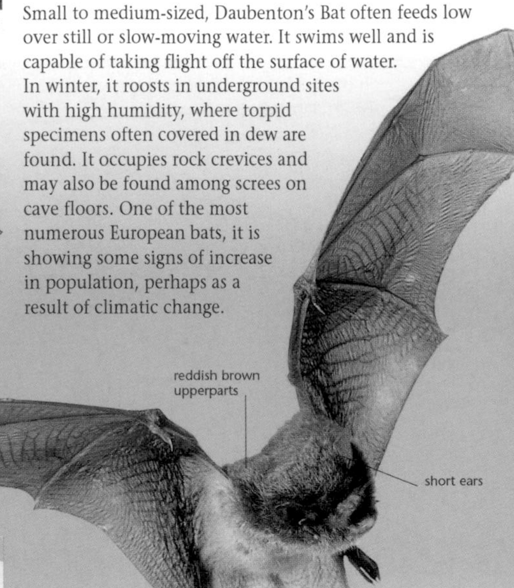


**NOTE**

*A small bat, flickering low over still water at a constant height looking for prey, is likely to be Daubenton's. However, while other species feed over water, this bat does not always do so.*

**SIZE** *Body 4.5–5.5cm; tail 3–3.5cm.*
**YOUNG** *Single young; June–July.*
**DIET** *Flying insects, such as caddisflies.*
**ECHOLOCATION** *45kHz.*
**STATUS** *Common.*
**SIMILAR SPECIES** *Pond and Long-fingered Bats (p.33), which are generally larger.*

# Pond Bat

*Myotis dasycneme* (Chiroptera)

Rather like a large, yellowish Daubenton's Bat, the migratory Pond Bat also prefers wetland habitats. It mostly feeds on insects taken low over water or picked from the surface in flight. In summer, it roosts in buildings or tree holes, while in winter it mostly roosts underground.

**FOUND** *typically in well-wooded areas, with easy access to still water; underground roosts are often in the foothills of mountains.*

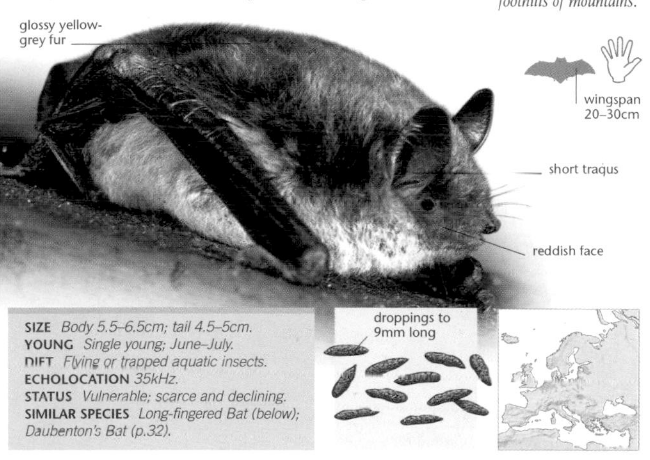


**SIZE** *Body 5.5–6.5cm; tail 4.5–5cm.*
**YOUNG** *Single young; June–July.*
**DIET** *Flying or trapped aquatic insects.*
**ECHOLOCATION** *35kHz.*
**STATUS** *Vulnerable; scarce and declining.*
**SIMILAR SPECIES** *Long-fingered Bat (below); Daubenton's Bat (p.32).*

droppings to 9mm long

# Long-fingered Bat

*Myotis capaccinii* (Chiroptera)

The Long-fingered Bat shares similar habitat preferences with Daubenton's, but is larger, greyer, and has a distinctly hairy tail membrane. It also feeds over water, often scooping insects from the surface with its large feet and tail. A gregarious species, it is often found together with several other types of bats.

**FORAGES** *mainly around water, usually in well-wooded areas, often on limestone; roosts underground at all seasons.*

pointed tragus

grey-brown fur with yellowish tints

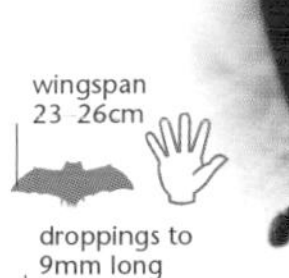


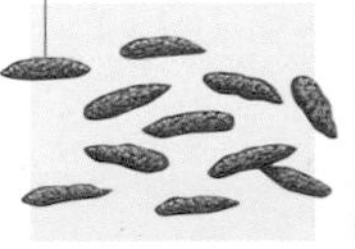

**SIZE** *Body 4.5–5.5cm; tail 3.5–4cm.*
**YOUNG** *Single young; June–July.*
**DIET** *Insects, flying and trapped on water.*
**ECHOLOCATION** *45kHz.*
**STATUS** *Vulnerable.*
**SIMILAR SPECIES** *Pond Bat (above); Daubenton's Bat (p.32).*

# Whiskered Bat

*Myotis mystacinus* (Chiroptera)

**INHABITS** *mixed landscape, with open woodland, gardens, and usually, waterbodies.*

Along with the similar and equally dumpy Brandt's Bat, the Whiskered Bat is the smallest of the widespread *Myotis* species. Both have very dark skin on the muzzle, ears, wings, and tail, and have such minor dissimilarities that they were not recognized as distinct until as recently as 1970. The Whiskered Bat emerges early in the evening and usually follows a regular hunting route. It roosts in trees and buildings in summer, but underground in winter.

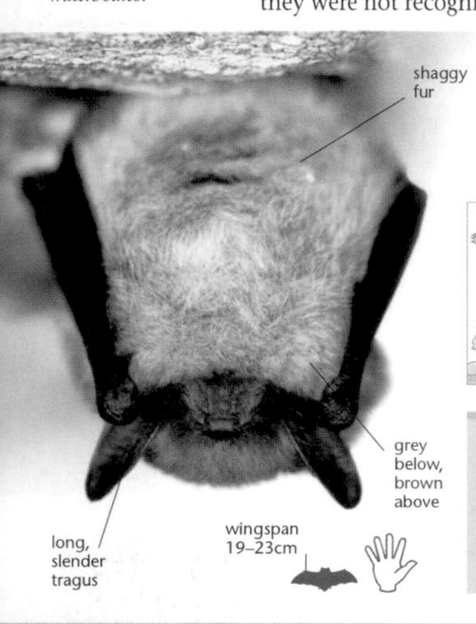

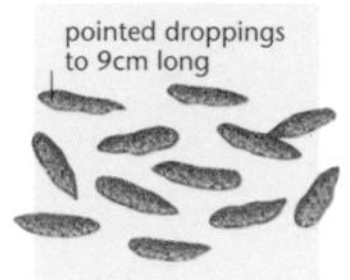


**SIZE** *Body 3.5–5cm; tail 3-4.5cm.*
**YOUNG** *Single young; June–July.*
**DIET** *Small flying insects, and spiders.*
**ECHOLOCATION** *60kHz.*
**STATUS** *Vulnerable; locally common.*
**SIMILAR SPECIES** *Brandt's Bat (below); Common Pipistrelle (p.42).*

# Brandt's Bat

*Myotis brandtii* (Chiroptera)

**OCCUPIES** *mixed lowland landscape, often in woodland; roosts in buildings in summer and caves in winter.*

This bat shares the diminutive size and dark skin of its close relative, the Whiskered Bat, but has more reddish brown fur, sometimes with a golden sheen. A definitive differentiation requires an examination of the teeth or the male genitalia, which is club-shaped in Brandt's Bat and thin in the Whiskered Bat.

**SIZE** *Body 3.5–5cm; tail 3.2-4.5cm.*
**YOUNG** *Single young; June–July.*
**DIET** *Flies and other small flying insects.*
**ECHOLOCATION** *60kHz.*
**STATUS** *Locally common.*
**SIMILAR SPECIES** *Whiskered Bat (above); Common Pipistrelle (p.42).*

# Greater Mouse-eared Bat

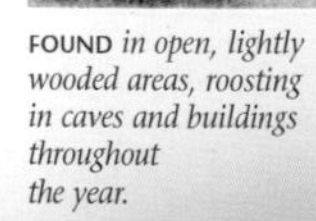

**FOUND** *in open, lightly wooded areas, roosting in caves and buildings throughout the year.*

*Myotis myotis* (Chiroptera)

A very large bat, the Greater Mouse-eared Bat differs from other large species by its pointed tragus, usually with a black spot near the tip. As befits its size, it has a slow, deliberate, direct flight action, foraging over open countryside and gardens at a height of up to 10m. It is highly colonial, especially in summer when several hundred females may gather in nursery roosts. Being a warmth-loving species, its northerly populations are migratory.

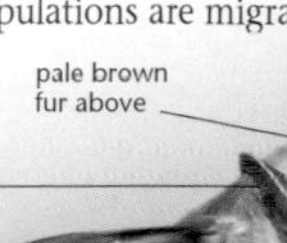


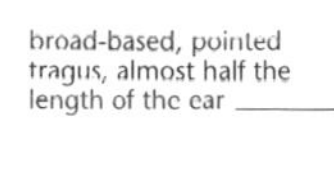


### NOTE

*In Britain, the Greater Mouse-eared Bat was first recorded in 1958, but this attempted colonization by the species died out in 1990. However, the discovery of a new individual in 2001 mirrors very recent population increases in parts of mainland Europe.*

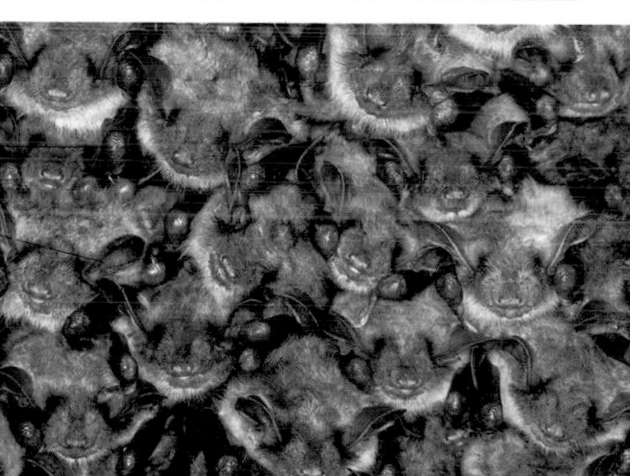


ROOSTING SITE

**SIZE** *Body 6.5–9cm; tail 4.5-6cm.*
**YOUNG** *Single young; born June–July.*
**DIET** *Large beetles, moths, and other insects.*
**ECHOLOCATION** *35kHz.*
**STATUS** *Near-threatened; locally common, but declining.*
**SIMILAR SPECIES** *Lesser Mouse-eared Bat (p.36), which has shorter, narrower ears;* M. punicus, *from Africa, on some Mediterranean islands.*

# Lesser Mouse-eared Bat

**OCCURS** *in open, warm, lightly wooded areas, roosting in caves, buildings, and occasionally, tree holes.*

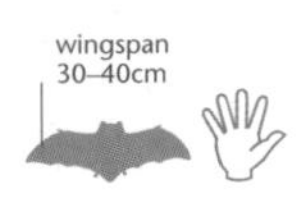

*Myotis blythii* (Chiroptera)

Despite its name, the Lesser Mouse-eared Bat is only marginally smaller than the Greater; indeed, some individuals overlap in size. Skull and teeth details may be needed to identify them, but its shorter ears, often with a whitish spot between them, and more graceful flight, usually clinches identification.

**SIZE** *Body 6–7.5cm; tail 4.5–6cm.*
**YOUNG** *Single young, June–July.*
**DIET** *Moths, beetles, and grasshoppers.*
**ECHOLOCATION** *35kHz.*
**STATUS** *Scarce and declining.*
**SIMILAR SPECIES** *Greater Mouse-eared Bat (p.35), which is slightly larger.*

# Bechstein's Bat

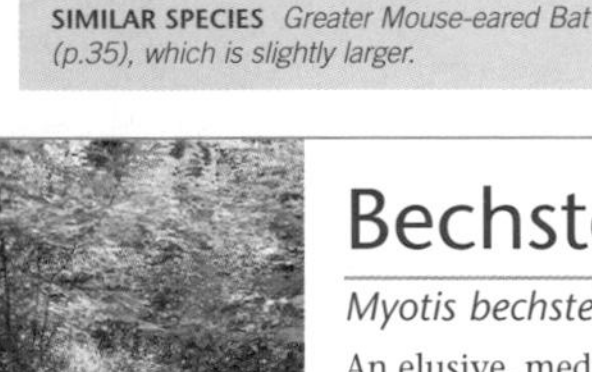

**FOUND** *in woodland, especially oak and beech, parks, and gardens; roosts in tree holes, caves, and buildings in winter.*

*Myotis bechsteinii* (Chiroptera)

An elusive, medium-sized species, Bechstein's Bat is closely associated with extensive deciduous woodland; unlike many of its close relatives, colony sizes are generally small. It has whitish underparts and a long, tapering tragus. Its noticeably long, shiny ears, with 9–11 creases, extend beyond the tip of the nose when they are folded forward.

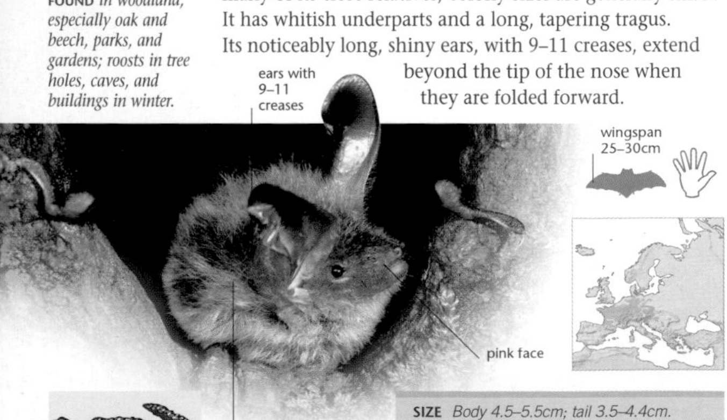


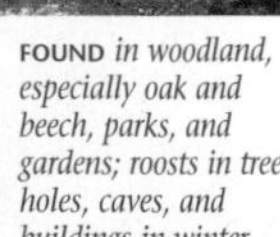


**SIZE** *Body 4.5–5.5cm; tail 3.5–4.4cm.*
**YOUNG** *Single young, June–July.*
**DIET** *Moths, flies, and other insects.*
**ECHOLOCATION** *50kHz.*
**STATUS** *Vulnerable; widespread but rare.*
**SIMILAR SPECIES** *Other small* Myotis *species; long-eared bats (pp.46–47).*

# Natterer's Bat

*Myotis nattereri* (Chiroptera)

Natterer's Bat emerges just after sunset, and flies low with a slow, controlled flight, often hovering. In calm conditions, it has a distinctive habit of flying with its tail pointed downwards. A medium-sized bat, it has a distinctly sinuous calcar (the bone supporting the tail membrane), and the rear margin of the tail has a dense fringe of hairs.

**INHABITS** *open woodland and farmland, especially near wet habitats.*

wingspan 24–30cm

pink-tinged membranes

long, slim tragus

white underparts

sinuous calcar

ROOSTING

**SIZE** *Body 4–5cm; tail 4–5cm.*
**YOUNG** *Single young, June–July.*
**DIET** *Insects (flies, caddisflies, and beetles).*
**ECHOLOCATION** *45kHz.*
**STATUS** *Scarce or rare, but widespread.*
**SIMILAR SPECIES** *Other medium-sized* Myotis *bats, but none have the sinuous calcar.*

droppings to 1.1cm long

# Geoffroy's Bat

*Myotis emarginatus* (Chiroptera)

Resembling a Natterer's Bat, the most distinctive feature of the shaggy Geoffroy's Bat is the sharp angle in the rear margin of its ear. The tip of its tail projects just beyond the end of the tail membrane and there is a sparse fringe of hairs on the tail margin. It is colonial, often roosting with other species, especially horseshoe-bats, and has now adopted urban habitats.

**FORAGES** *in woodland and scrubby habitats, and, increasingly, in urban areas.*

reddish, shaggy fur

wingspan 22–25cm

droppings to 1.2cm long

angled ear margin

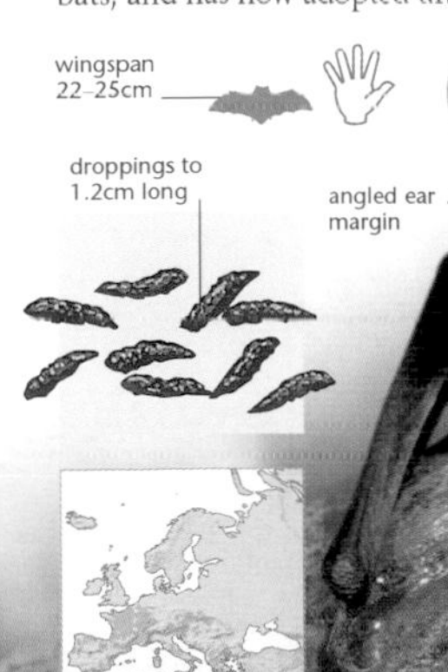

**SIZE** *Body 4–5.3cm; tail 3.8–4.5cm.*
**YOUNG** *Single young, June–July.*
**DIET** *Spiders and other invertebrates.*
**ECHOLOCATION** *50kHz.*
**STATUS** *Vulnerable; very locally common.*
**SIMILAR SPECIES** *Other medium-sized* Myotis *bats, but none has the angled ear margin.*

# Noctule

*Nyctalus noctula* (Chiroptera)

**FEEDS** *over open woodland, pasture, and urban parks; summer roosts mostly in tree-holes, while winter roosts, often very large, in trees and rock crevices.*

One of the largest widespread European bats, Noctules are relatively visible as a result of their early emergence from roosts, up to an hour before dark. In such conditions, the distinctive red-brown colour of their upperparts can be clearly seen. Typically, they fly high (at 50m or more), but it is not difficult to find them even in the dark because the echolocation calls can be picked up on a bat detector at a range of 200m. Noctules also produce an audible (to most people) loud metallic chirp in flight.

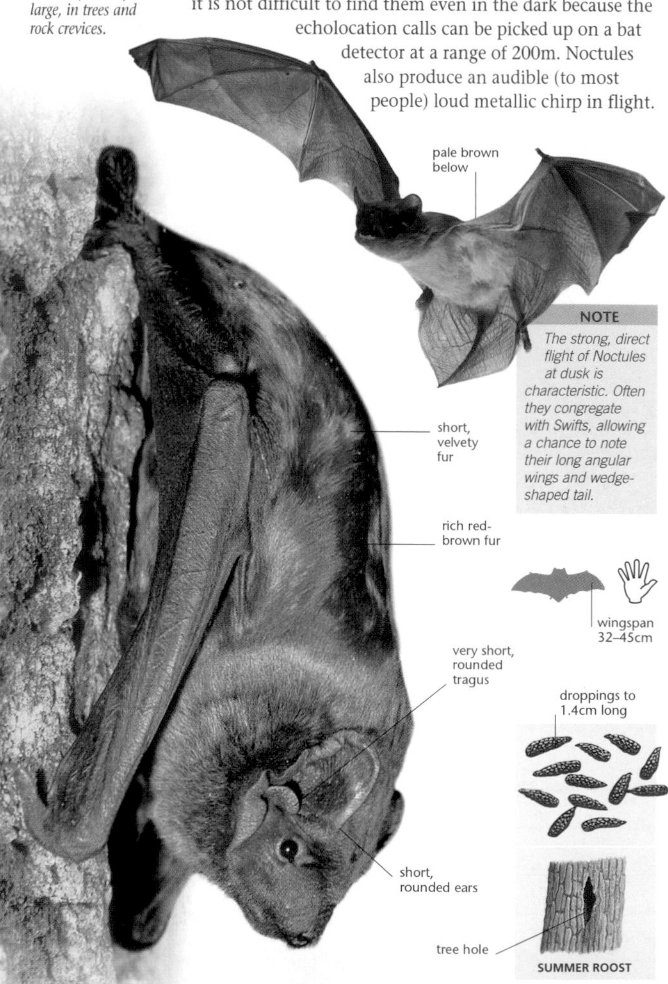


**NOTE**

*The strong, direct flight of Noctules at dusk is characteristic. Often they congregate with Swifts, allowing a chance to note their long angular wings and wedge-shaped tail.*

**SIZE** *Body 6–8cm; tail 4–6cm.*
**YOUNG** *Single litter of 1–3; June–July.*
**DIET** *Large flying insects, including moths and beetles.*
**ECHOLOCATION** *25kHz.*
**STATUS** *Common, some local declines; migratory in the east.*
**SIMILAR SPECIES** *Leisler's Bat (p.39); Serotine (p.40); Greater Noctule* (N. lastopterus), *which is larger and much rarer.*

# Leisler's Bat

*Nyctalus leisleri* (Chiroptera)

A medium-sized bat, with long, narrow wings, Leisler's Bat is generally darker than similar species, such as the Noctule (p.38), its shaggy, red-tinged dark brown fur, being made up of two-tone hair, is darker at the roots than the tip. The ears and tragus are both short and rounded, and the dark brown wing membranes have a sparse but distinct covering of hair beneath. Largely a migratory species, individual Leisler's Bats have been reported as moving more than a thousand kilometres between breeding and wintering grounds. On migration, they are frequently seen flying by day.

**FORAGES** *after sunset over open lowland areas, with scattered trees and woodland; roosts mainly in tree-holes.*

short ears and tragus

pale brown below

**NOTE**

*Ireland constitutes the world stronghold of the Leisler's Bat, where it thrives in the absence of other large bats. Here, most nursery roosts are found in roof spaces.*

red-brown fur, with dark roots

droppings to 0.9cm long

wingspan 30–34cm

**SIZE** *Body 5–6.8cm; tail 3.5–5cm.*
**YOUNG** *Single litter of one or two; June–July.*
**DIET** *Flying insects, especially beetles, moths, and caddisflies.*
**ECHOLOCATION** *25kHz.*
**STATUS** *Near-threatened; generally scarce but common in Ireland.*
**SIMILAR SPECIES** *Noctule (p.38), which is larger and redder; Schreiber's Bat (p.44); Azorean Bat* (N. azoreum)*, which hunts by day.*

# Serotine

*Eptesicus serotinus* (Chiroptera)

**FAVOURS** *open and lightly wooded habitats in the lowlands; roosts in buildings, occasionally tree-holes, and in winter may be found underground.*

One of the larger bats, the Serotine is often seen feeding at dusk with Swifts, though at a somewhat lower level than Noctules (p.38). In reasonable light, the dark upperparts can be seen to contrast with the more yellowish underparts. At rest, the very dark brown wing membranes, and especially its face, are characteristic, together with its short, rounded tragus (the front edge is concave and the rear edge convex). It has a protruding tail-tip, and large, powerful teeth, necessary for tackling hard-bodied beetles. In parts of its range, including southern Britain, it is one of the more abundant species, and associated with buildings.

coarse droppings to 1cm long

wingspan 32–38cm

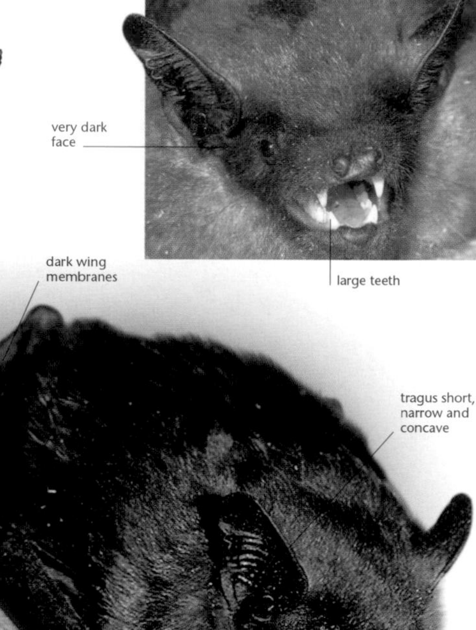


**NOTE**

*The summer roosts of Serotines in buildings are very obvious: just before sunset, the occupants gather inside the entrance hole, with a noisy, high-pitched chattering.*

**SIZE** *Body 6–8cm; tail 4.5–5.5cm.*
**YOUNG** *Single young; June–August.*
**DIET** *Large flying insects, especially dung beetles and moths.*
**ECHOLOCATION** *25kHz.*
**STATUS** *Locally common; perhaps expanding its range.*
**SIMILAR SPECIES** *Noctule (p.38) and mouse-eared bats (pp.35–36); Botta's Serotine* (E. bottae)*, which is restricted to Rhodes.*

# Northern Bat

*Eptesicus nilssonii* (Chiroptera)

The most northerly distributed bat in Europe, the Northern Bat is of a medium size, and has a rapid, agile flight; it is often seen flying during daylight hours. Its very dark face and ears, and projecting tail-tip indicate a close relationship to the Serotine, but the pale brown underside and sandy hair-tips on the back are distinctive.

**INHABITS** *largely forested areas, roosting especially in buildings, tree-holes, and caves.*

shiny blackish wing membranes

blackish face and ears

tapered droppings to 1cm long

**SIZE** *Body 5–7cm; tail 3.5–4cm;*
**YOUNG** *Single litter of 1–2; June–July.*
**DIET** *Small flies and other insects.*
**ECHOLOCATION** *30kHz; loud chirrups.*
**STATUS** *Common in northern areas.*
**SIMILAR SPECIES** *Parti-coloured Bat (below), which appears frosted; Savi's Pipistrelle (p.44)*

wingspan 27cm

# Parti-coloured Bat

*Vespertilio murinus* (Chiroptera)

The Parti-coloured Bat is one of the most distinctive medium-sized European bats, with its almost white underparts, sharply divided from the dark upper fur, and the frosted appearance of the upperparts. Unfortunately, these features are rarely seen in flight as it flies exclusively at night.

**FOUND** *around woodland, farmland, and urban areas.*

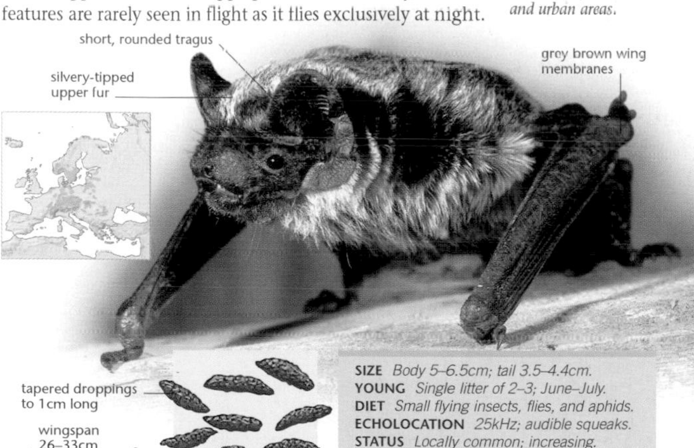


**SIZE** *Body 5–6.5cm; tail 3.5–4.4cm.*
**YOUNG** *Single litter of 2–3; June–July.*
**DIET** *Small flying insects, flies, and aphids.*
**ECHOLOCATION** *25kHz; audible squeaks.*
**STATUS** *Locally common; increasing.*
**SIMILAR SPECIES** *Northern Bat (above), which does not appear frosted.*

**OCCURS** *in all lowland habitats, including woodland, heathland, farmland, and urban areas, especially around water; roosts primarily in buildings.*

# Common Pipistrelle

*Pipistrellus pipistrellus* (Chiroptera)

The commonest bat over most of Europe, it is also the smallest. This strongly colonial bat is a rather variable brown above, pale below, but each individual is generally uniform in colour. As with other pipistrelles, the wing membrane extends outside the calcar, but the thumb is relatively short. It is quite common to find maternity roosts of more than a thousand females. Its hunting flight is rather jerky, usually below 10m in height. Activity continues later into the autumn than most other bats and midwinter emergence is not uncommon.

wingspan 18–24cm

short, rounded ears

SOPRANO PIPISTRELLE

### NOTE

*The Common Pipistrelle has recently been recognized as two species, the second being the Soprano Pipistrelle (P. pygmaeus), so-called for its higher echolocation frequency of 55kHz. Otherwise, there are only minor visible differences, and perhaps distributional and ecological ones.*

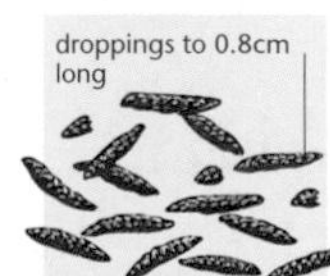


**SIZE** *Body 3.3–5.2cm; tail 2.5–3.6cm.*
**YOUNG** *Single litter of 1–2; May–August; mostly breed every other year.*
**DIET** *Small flying insects: midges, caddisflies, lacewings, and moths.*
**STATUS** *Common, though declining.*
**ECHOLOCATION** *45kHz.*
**SIMILAR SPECIES** *Other pipistrelles (pp.43–44); Whiskered Bat (p.34); Brandt's Bat (p.34), both of which have narrow ears and tragus.*

# Nathusius' Pipistrelle

*Pipistrellus nathusii* (Chiroptera)

**FAVOURS** *woodland; summer roosts in tree holes; also on walls and in caves in winter.*

Rather similar to the Common Pipistrelle, Nathusius' Pipistrelle is a little larger and much less frequently associated with human habitats. Its upperparts darken in the winter and lack the uniform appearance of related species. On close examination, the hairs on its tail, large teeth, and long thumb are characteristic. Highly migratory, it is occasionally recorded as a vagrant, far from its core areas.

cylindrical droppings 0.8cm long

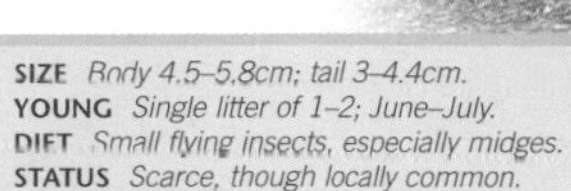

**SIZE** *Body 4.5–5.8cm; tail 3–4.4cm.*
**YOUNG** *Single litter of 1–2; June–July.*
**DIET** *Small flying insects, especially midges.*
**STATUS** *Scarce, though locally common.*
**ECHOLOCATION** *35kHz.*
**SIMILAR SPECIES** *Common Pipistrelle, (p.42), which is smaller.*

wingspan 22–25cm

# Kuhl's Pipistrelle

*Pipistrellus kuhlii* (Chiroptera)

Generally southern in distribution, Kuhl's Pipistrelle is closely associated with human habitation. Generally paler brown, especially below, than Common and Nathusius' Pipistrelles, its most distinctive feature is the whitish rear edge of the wing membrane. It often hunts in groups around street-lamps.

**FREQUENTS** *urban areas; summer roosts in buildings and winter roosts in cellars and rock crevices.*

**IN FLIGHT**

whitish edge

yellowish brown fur above

pale below

droppings to 0.8cm

**SIZE** *Body 4–4.7cm; tail 3–4.5cm.*
**YOUNG** *Single litter of 1–2; June–July.*
**DIET** *Small flying insects.*
**STATUS** *Common.*
**ECHOLOCATION** *40kHz.*
**SIMILAR SPECIES** *Other pipistrelles (pp.42–44);* Pipistrellus maderensis.

wingspan 20–24cm

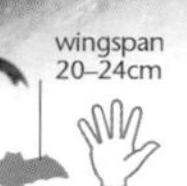

# Savi's Pipistrelle

*Pipistrellus savii* (Chiroptera)

**FOUND** *in rocky valleys, extending into the mountains; roosts in rock crevices, and occasionally, in buildings and trees.*

Recent research suggests that Savi's Pipistrelle, a poorly known southern species, should perhaps not be grouped with the pipistrelles but should be known as *Hypsugo savii*. Its broad, rounded ears, together with frosted brown fur on its upperparts that sharply contrasts with off-white underparts, are distinctive. Its relatively slow, direct flight, often above roof-top height, contrasts with the lower, more erratic flight typical of pipistrelles.

frosted brown upperparts

long, shaggy fur

droppings to 0.8cm long

off-white below

wingspan 22–25cm

**SIZE** *Body 4–5.4cm; tail 3-4cm.*
**YOUNG** *Single litter of 1–2; June–July.*
**DIET** *Small flying insects.*
**ECHOLOCATION** *35kHz.*
**STATUS** *Scarce, but locally common, and perhaps overlooked.*
**SIMILAR SPECIES** *Pipistrelles (pp.42–43).*

# Schreiber's Bat

*Miniopterus schreibersii* (Chiroptera)

**OCCURS** *in open, rocky areas, especially on limestone, where caves often host very large roosts.*

Schreiber's Bat is the European bat with the widest world distribution, extending across Africa, Asia, and Australasia. Medium-sized, it has very long, narrow wings, which give it a fast, almost swallow-like flight. The fur is a dark grey-brown, slightly paler below, and, especially on the head, short, velvety, and erect.

short, erect fur on head

smoky grey-brown fur

droppings to 0.9cm long

wingspan 30–34cm

**SIZE** *Body 5–6.2cm; tail 5.5–6.5cm.*
**YOUNG** *Single young; June–July.*
**DIET** *Small flying insects: flies and moths.*
**ECHOLOCATION** *55kHz.*
**STATUS** *Near-threatened; locally common, but at risk because it is highly colonial.*
**SIMILAR SPECIES** *Leisler's Bat (p.39).*

# Barbastelle

*Barbastella barbastellus* (Chiroptera)

A medium-sized bat, the Barbastelle is very distinctive in its overall dark appearance: the face and ears are black, the wings dark brown, and the fur blackish brown (a little paler below). However, silvery hair-tips give it a somewhat frosted look. It is also identifiable by its characteristic blunt, pug-like face. The ears are moderately long, to 1.8cm, but wide, joined on top of the head, and the tragus is long and triangular. Populations are very dispersed, especially in summer, and as a result its breeding habits are poorly known. Winter cave roosts in eastern Europe can hold several hundred individuals.

**LIVES** *primarily in woodland habitats; feeding in woodland rides, over water, and in gardens (montane and submontane in some areas); roosts in trees and buildings, and caves in winter.*

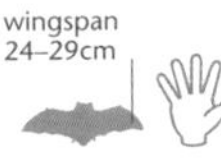


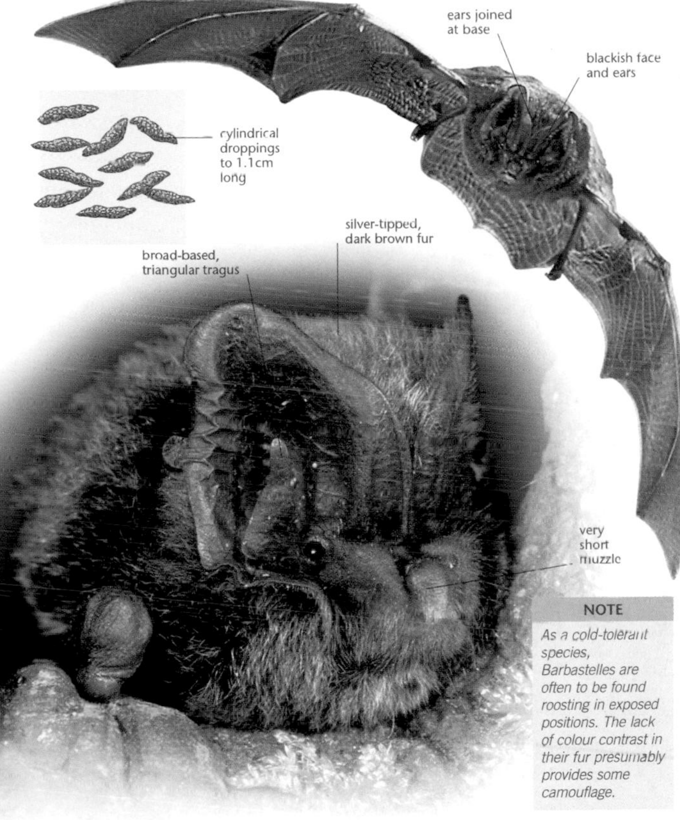


**NOTE**

*As a cold-tolerant species, Barbastelles are often to be found roosting in exposed positions. The lack of colour contrast in their fur presumably provides some camouflage.*

**SIZE** *Body 4.5–5.8cm; tail 4–5.4cm.*
**YOUNG** *Single young (occasionally two); June–July.*
**DIET** *Small flies, moths, and other soft-bodied invertebrates; it's weak jaws cannot tackle hard-shelled insects.*
**ECHOLOCATION** *30kHz.*
**STATUS** *Vulnerable; rare in the west, locally common further east.*
**SIMILAR SPECIES** *None.*

# Brown Long-eared Bat

*Plecotus auritus* (Chiroptera)

**PREFERS** *well-wooded, open areas, such as parks and gardens; roosts in buildings and trees throughout the year, and occasionally underground in winter.*

Very long ears, up to 4cm long, serve to distinguish the long-eared bats from all other species. At rest, the ears are folded back over the body, although the long, narrowly triangular tragus remains erect. Its thumb is more than 6mm long. The ears and face of the Brown Long-eared Bat are pinkish brown, while the fur on its upperparts is pale brown, fading to whitish below. Its flight is graceful and swooping, often with well-controlled hovering as it gleans insects off the foliage of trees.

thumb longer than 6mm

brown above, paler below

fluffy fur

very long, pinkish ears

droppings to 1cm long

discarded moth wings

FEEDING PILE

long tapering tragus, 5mm or less wide

wingspan 24–29cm

small pinkish brown face

RESTING

### NOTE

*This bat generally takes its captured prey back to a feeding perch before removing the inedible parts: look out for piles of moth wings, often in barns.*

**SIZE** *Body 3.5–5.3cm; tail 3.5–5.5cm.*
**YOUNG** *Single young (occasionally two); June–July.*
**DIET** *Flying insects, especially moths; spiders and other invertebrates.*
**ECHOLOCATION** *50kHz, though only very quietly.*
**STATUS** *Common.*
**SIMILAR SPECIES** *Grey Long-eared Bat (p.47); Tenerife Long-eared Bat* (P. teneriffae), *which is found only on the Canary Islands.*

# Grey Long-eared Bat

*Plecotus austriacus* (Chiroptera)

Rather similar in overall appearance to the Brown Long-eared Bat, the slightly larger Grey Long-eared Bat, if seen clearly, is usually distinctly grey-brown, especially on the face and ears. Even the whitish underparts appear slate-grey when the fur is parted, as the hair has dark roots. Otherwise, the tragus is wider and more opaque, and the thumb is short, less than 6mm long. Even more difficult to see, but a defining characteristic, is the club-shaped penis of the male. Roosting bat are often to be seen hanging free on walls rather than wedged into cracks.

**OCCUPIES** *open and agricultural habitats, often feeding around streetlights; roosts mainly in buildings and rock cavities.*

tragus 6mm wide

short thumb

whitish below

dark grey face

broad wings with blackish membranes

knobbly droppings to 1.1cm long

wingspan 25–35cm

**NOTE**

*Montane long-eared bats, with features intermediate between Common and Grey are recently thought to represent a new species* P. alpinus.

**SIZE** *Body 4–5.8cm; tail 3.5–5.5cm.*
**YOUNG** *Single young; June.*
**DIET** *Small to medium-sized flying insects, especially moths.*
**ECHOLOCATION** *50kHz.*
**STATUS** *Scarce, to common in the south.*
**SIMILAR SPECIES** *Brown Long-eared Bat (p.46); Tenerife Long-eared Bat* (P. teneriffae), *which is larger and darker.*

**FREQUENTS** *rocky valleys, gorges, and high mountains, often found around human habitation; roosts in crevices, caves, and buildings.*

# European Free-tailed Bat

*Tadarida teniotis* (Chiroptera)

A very large bat, the European Free-tailed Bat is unique among European species in that a large proportion (up to a half) of its tail projects beyond the flight membrane. It has large, rounded ears, which meet at their base, and are held projecting forwards in flight; the face, ears (without a tragus), and wings are all dark grey. The fur is soft with a velvety texture. Although it hibernates, it is frequently active during the winter, perhaps a result of its southerly range.

**NOTE**

*European Free-tailed Bats are high and fast fliers, often associating with swifts, and easily detected even at night by their loud, sharp, audible chirp as they hurtle around above rooftop height.*

long, narrow wings

dark grey-brown above

wingspan 45cm

long tail

pale grey below

forward-pointing ears

dark grey face

cylindrical droppings to 1.8cm long

short fur

projecting lobe on outer edge of ear

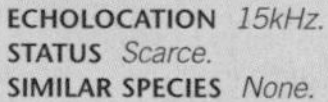

**SIZE** *Body 8–9cm; tail 4.5–6cm.*
**YOUNG** *Single young; born in summer.*
**DIET** *Flying insects; moves to mountain passes to feed on migrating species.*
**ECHOLOCATION** *15kHz.*
**STATUS** *Scarce.*
**SIMILAR SPECIES** *None.*

# Barbary Ape

*Macaca sylvanus* (Primates)

The only wild primate in Europe, the Barbary Ape is one of only a very few tail-less monkey species. In the European context, it is unmistakable, with its shaggy brown fur, deep brow, and opposable thumbs. The current population on the Rock of Gibraltar is probably derived from introductions.

**OCCURS** *only on dry, rocky, scrubby hillsides on the Rock of Gibraltar.*

hand to 12cm

foot to 17cm

segmented droppings to 8cm

**SIZE** *Body 60–70cm; shoulder height 45cm.*
**YOUNG** *Single young; born in summer.*
**DIET** *Omnivorous: feeds on fruit, leaves, bark, insects, reptiles, and small mammals.*
**STATUS** *Vulnerable; six troops numbering some 150 animals.*
**SIMILAR SPECIES** *None.*

# Red-necked Wallaby

*Macropus rufogriseus* (Marsupialia)

The only wild marsupial in Europe, Red-necked Wallabies are easily identified. The upright, bipedal, bounding gait is very different from the loping of the similarly-sized Brown Hare (p.51), the only mammal with which it may possibly be confused. Its black feet and tail-tip are also particularly noticeable.

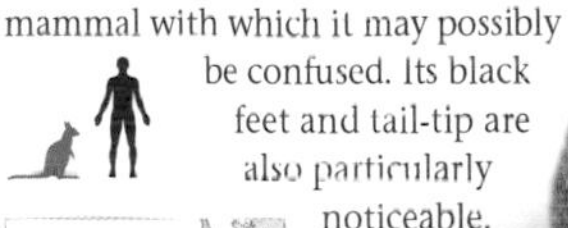

**FAVOURS** *dense scrub or bracken during the day; not strictly nocturnal.*

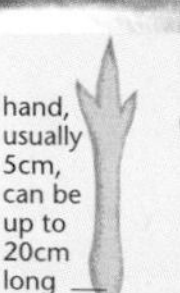

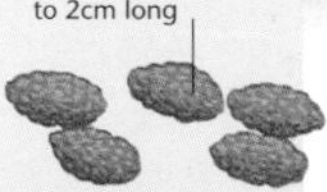


**SIZE** *Body 60–70cm; tail 60–75cm.*
**YOUNG** *Single young; born March–May.*
**DIET** *Browses on heather and related shrubs, bracken, and pine needles.*
**STATUS** *Introduced from Australia; tentatively established in Britain since about 1940.*
**SIMILAR SPECIES** *None.*

# Mountain Hare

*Lepus timidus* (Lagomorpha)

**FAVOURS** *montane grassland, woodland, and tundra; sometimes found in lowland. Unlike other hares, it makes short burrows.*

Also known as Blue Hare (due to its thick, insulating undercoat of dark-blue grey fur), the Mountain Hare's coat turns white in winter, and its feet are thick with fur. In lowland and agricultural areas, the hare has a brownish coat which does not turn white; the relatively short ears and wholly white tail are characteristic features to look out for.

WINTER COAT

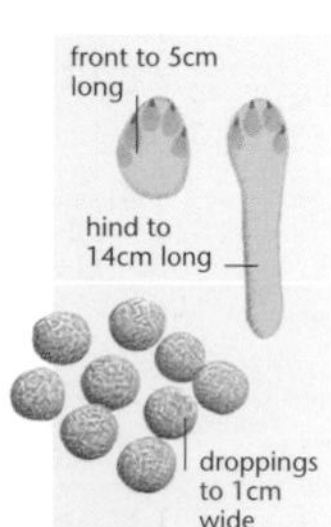

**SIZE** *Body 45–60cm; tail 4–8cm.*
**YOUNG** *Two or three litters of up to five.*
**DIET** *Grasses, herbs, low shrubs, and bark.*
**STATUS** *Common in suitable habitats; population density fluctuates in the north.*
**SIMILAR SPECIES** *Iberian Hare (below), which has larger ears; Brown Hare (p.51).*

# Iberian Hare

*Lepus granatensis* (Lagomorpha)

**INHABITS** *open agricultural and cultivated land, sand dunes, woodland and mountain areas.*

The south-western counterpart of the Brown Hare, Iberian Hares are generally smaller, and show a distinct rufous tinge to the fur, especially on the thighs and flanks. The true affinities of Iberian Hares have been extensively debated, some considering them to be a form of Brown Hare, and others a form of the widespread African species, the Cape Hare.

front print to 3cm long

hind to 12cm long

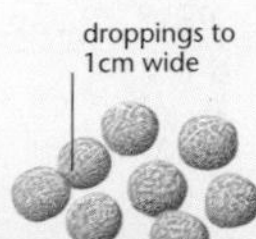


**SIZE** *Body 45–54cm; tail 9–11cm.*
**YOUNG** *Three or four litters of up to seven.*
**DIET** *Cereals, grasses, and strips of bark.*
**STATUS** *Locally common.*
**SIMILAR SPECIES** *Mountain Hare (above) Brown Hare (p.51); Broom Hare* (L. castroviejoi) *in N.W. Iberian Peninsula.*

# Brown Hare

*Lepus europaeus* (Lagomorpha)

A widespread European species, Brown Hares have very long ears, and large, powerful legs which can generate speeds of up to 75km per hour for short bursts. Unlike a Rabbit, with which a young, short-eared specimen may be confused, the tail is held depressed when running, so that the rump does not show as clear white, and the ears are tipped with black. Their activity takes place mostly at dusk and dawn, but they can be active at any time, especially during the springtime display period. Field signs include flattened patches of grass, or forms, where they rest by day and rear their young.

**FREQUENTS** *open agricultural fields and pastures, also hedgerows and woodland.*

**LEVERET**

**NOTE**

*"Mad March Hares" that are often seen chasing and boxing around open fields in the spring, are actually the slightly larger females deterring the unwanted advances of males.*

black tip

ears more than 8cm long

rich brown fur, reddening in winter

tail blackish above

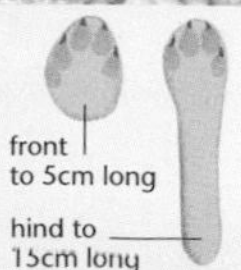


droppings
1cm wide

flattened grass

**FORM**

bark
stripping

**SAPLING DAMAGE**

**SIZE** *Body 50–70cm; tail 7–10cm.*
**YOUNG** *About three litters of up to four; February–October.*
**DIET** *Grazes herbs, cereals, and grasses; browses low shrubs and strips bark.*
**STATUS** *Common, though declining in parts.*
**SIMILAR SPECIES** *Rabbit (p.52); other hares (p.50); Corsican Hare* (L. corsicanus), *which is smaller and restricted to S. Italy.*

# Rabbit

*Oryctolagus cuniculus* (Lagomorpho)

**FREQUENTS** *grassland and farmland, with hedges; look out for signs such as burrows, scrapes, and droppings, placed communally in latrines.*

An ancient introduction for food and fur to much of Europe from its original home in Iberia, the Rabbit is now a familiar part of the European countryside, one which exerts a considerable influence on agriculture and natural habitats through grazing and other activities. Its fur is generally grey-brown above and paler below, but a wide range of colour variations from black to white can persist in the wild after escapes from captivity. The legs are relatively short, yet powerful, and the ears are long, though shorter than those of hares, and lack black tips. The tail is brown above and white below; when running, the tail is held erect to display the white fur to advantage, as a warning of danger to the social group.

powerful hind legs

WARREN

long ears

large bulbous eyes

rounded muzzle

**NOTE**

*"Breeding like a rabbit" is an apt phrase; most does are pregnant again within two days of giving birth: young born early in the season can breed the same year.*

front print to 3cm long

hind print to 8cm long

rounded droppings to 1cm wide

used for territorial marking

LATRINE

**SIZE** *Body 35–45cm; tail 4–8cm.*
**YOUNG** *Three to seven litters of up to 12 young; February–August.*
**DIET** *Grass, leaves, bulbs, and roots; strips bark in the winter.*
**STATUS** *Introduced to most of its range; common.*
**SIMILAR SPECIES** *Hares (pp.50–51), which have longer legs and ears, and a more upright stance; Eastern Cottontail* (S. floridanus)*, which has pale feet and a zig-zag run, and has been introduced for sport to N. Italy.*

# Red Squirrel

*Sciurus vulgaris* (Rodentia)

A remarkably variable species over much of its range, the Red Squirrel can be any colour from red, through all shades of brown, to black, though always with white underparts. In winter, all variations assume a more greyish appearance, and develop prominent ear-tufts. Red Squirrels are very agile, and spend much of their time in the branches of trees, where the long bushy tail is used for balance. However, a considerable amount of foraging is done on the ground, where they move with a light, bounding gait.

**FOUND** *in deciduous and coniferous woodland, where its nests (dreys) are built of twigs and leaves in the fork of a tree, normally close to the trunk.*

DARK FORM

long bushy tail

ear-tufts

white underparts

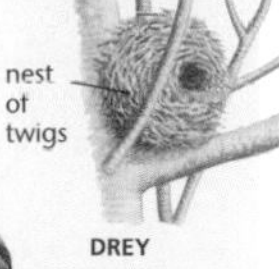


nest of twigs

DREY

scales removed whole

EATEN CONES

EATEN NUTS

**NOTE**

*One of the favoured foods of the Red Squirrel is pine seeds. A cone is held between the forefeet, and the scales are then gnawed off to reveal the seeds. The core is then dropped, providing one of the most obvious field signs of Red Squirrels.*

**SIZE** *Body 18–25cm; tail 24–20cm.*
**YOUNG** *One or two litters of 3–5; March–September.*
**DIET** *Tree seeds, buds, bark, roots, fungi, birds' eggs, and nestlings.*
**STATUS** *Near-threatened; common, but declining where it overlaps with Grey Squirrels.*
**SIMILAR SPECIES** *Grey Squirrel (p.54); Persian Squirrel* (S. anomalus), *which lacks ear tufts and has off-white underparts.*

# Grey Squirrel

*Sciurus carolinensis* (Rodentia)

**INHABITS** *all kinds of woodland, parkland, and gardens; where its confiding nature ensures a ready supply of food.*

Since its introduction in the 19th century, the robust Grey Squirrel has displaced the Red Squirrel from most of England and Wales, and a similar process now seems to be under way in northern Italy. Grey Squirrels are less arboreal than Red Squirrels, feeding extensively on the ground, and hiding caches of acorns and other fruit for future use. The grey fur is somewhat variable in colour, with reddish tones especially in juveniles, but the ears are never strongly tufted. Field signs are almost the same as for Red Squirrels, but the dreys are usually placed away from the main trunk of a tree.

DARK FORM

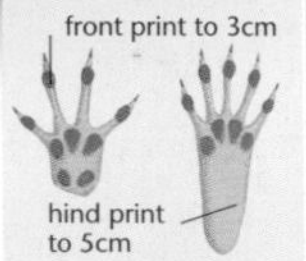

DREY

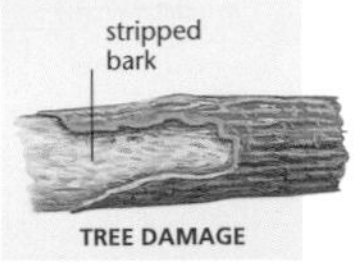


TREE DAMAGE

**SIZE** *Body 23–30cm; tail 20–24cm.*
**YOUNG** *One or two litters of 3–6; May–October.*
**DIET** *Seeds and other plant material; insects, birds' eggs, and nestlings.*
**STATUS** *Introduced from North America; very common and still extending its range.*
**SIMILAR SPECIES** *Greys from the brighter end of the spectrum can look like Red Squirrel (p.53); Fat Dormouse (p.74), which is smaller.*

# Flying Squirrel

*Pteromys volans* (Rodentia)

A nocturnal, arboreal animal, the Flying Squirrel has a greatly extended flap of skin between the ankle and wrist, which, when expanded, enables it to glide between trees. Just before landing, the tail is raised and the gliding membrane arched to act as a brake. Although characteristic of cold forest regions, it does not hibernate. Instead, it stores seeds and catkins in tree-holes to provide food for winter.

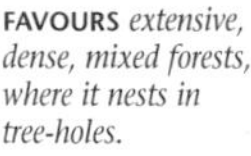

**FAVOURS** *extensive, dense, mixed forests, where it nests in tree-holes.*

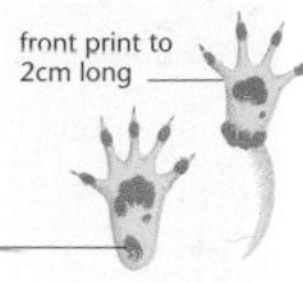


**SIZE** *Body 14–20cm; tail 9–14cm.*
**YOUNG** *Two litters of 2–4; April–July.*
**DIET** *Tree catkins, seeds, buds, and leaves; occasionally birds' eggs and nestlings.*
**STATUS** *Near-threatened; scarce and declining.*
**SIMILAR SPECIES** *Fat Dormouse (p.74).*

# Barbary Ground-squirrel

*Alantoxerus getulus* (Rodentia)

A single pair of Barbary Ground-squirrels was introduced to Fuerteventura (Canary Islands) in 1965 from North Africa. Its numbers grew rapidly, although the increase in population has recently stabilized at a moderate level. Surprisingly, there is no suggestion of severe economic or ecological impacts, although all steps are being taken to prevent its spread. A ground-dwelling species, its fur is marked by five buff stripes.

**OCCUPIES** *dry, open habitats, including agricultural areas, where there are rocks, scrub, or stone walls to provide refuge.*

small ears

grey-brown fur

buff back stripes

**SIZE** *Body up to 15cm; tail up to 15cm.*
**YOUNG** *Two litters of 4–6; February–August.*
**DIET** *Seeds, fruit, snails, carrion, and, reportedly, goat excrement.*
**STATUS** *Introduced; locally common.*
**SIMILAR SPECIES** *Siberian Chipmunk (p.58), which has large ears and stronger markings.*

# European Souslik

*Spermophilus citellus* (Rodentia)

**OCCUPIES** *dry grassland and cultivated areas, from sea level to the sub-alpine zone, in extensive warrens.*

The cylindrical body and short ears, legs, and tail of the European Souslik are ideally adapted to a largely subterranean life. Individual burrows are generally aggregated into large colonies, around which animals may be seen watching for predators. Hibernation takes place between October and March.

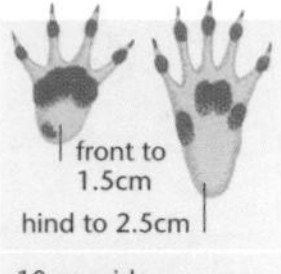


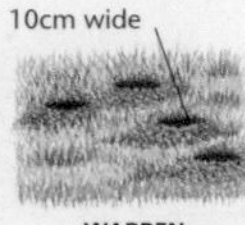

**SIZE** *Body 19–22cm; tail 6–7cm.*
**YOUNG** *Single litter of up to six; May–June.*
**DIET** *Largely seeds; other plant parts and insects.*
**STATUS** *Vulnerable; locally common but declining through agricultural changes.*
**SIMILAR SPECIES** *Spotted Souslik (below).*

# Spotted Souslik

*Spermophilus suslicus* (Rodentia)

**LIVES** *in dry grassland; also found in cultivated fields.*

Similar in size and shape to the European Souslik, the Spotted Souslik is easily differentiated by the creamy white spots on its darker brown back; otherwise, its tail is shorter and less bushy, and its European distribution much more restricted. Spotted Sousliks also live in dense aggregations in dry grassland, but they are seemingly more tolerant of ploughing, and are often pests of cultivated fields.

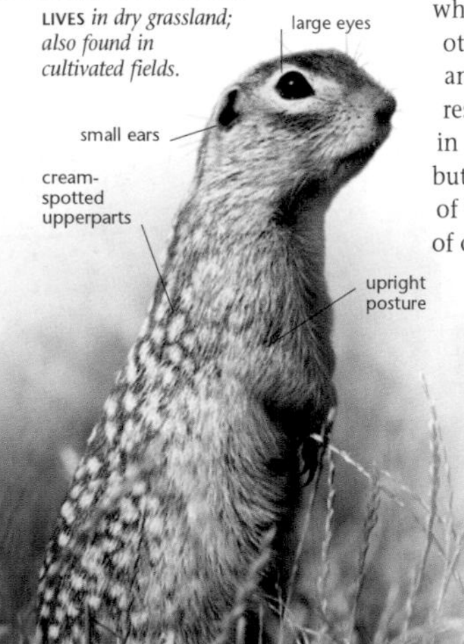

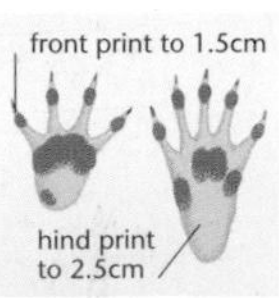


**SIZE** *Body 18–25cm; tail 3–4cm.*
**YOUNG** *Single litter of 4–8; May–June.*
**DIET** *Seeds and other plant material; ground-dwelling insects and other invertebrates.*
**STATUS** *Vulnerable; locally common, but many declines and local extinctions.*
**SIMILAR SPECIES** *European Souslik (above).*

# Alpine Marmot

*Marmota marmota* (Rodentia)

The Alpine Marmot is a large rodent with a moderately long bushy tail and short legs. They are often seen in open alpine habitats, sitting alert on a rock on the skyline, or spreadeagled in a sun-trap. Highly communal, they hibernate from October to April in the deeper parts of their warren system. Their dense fur and accumulated fat provide effective insulation from the cold, but also prevent cooling when it is hot: marmots then often retreat into their burrows during the heat of the day.

**FOUND** *in mountain pastures, usually between 1,000 and 3,000m, especially with rocky outcrops as vantage points.*

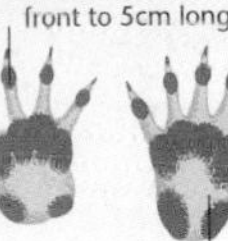


droppings to 6cm long

hole 20cm wide

**COMMUNAL BURROWS**

**NOTE**

*Often the first sign of this species is a loud, far-carrying, shrill whistle, more like a bird than a mammal, produced in alarm and for territorial purposes. If you hear it, scan prominent rocks in the vicinity to see the usually alert marmot.*

**SIZE** *Body 50–55cm; tail 15–20cm.*
**YOUNG** *Single litter of 2–6; May–June.*
**DIET** *Grass and other vegetation, including roots.*
**STATUS** *Locally common in core areas, but threatened in some outlying parts of its range; introduced in the Pyrenees and elsewhere, as a source of food, fur, and fat, the latter believed to have medicinal properties.*
**SIMILAR SPECIES** *None.*

# Siberian Chipmunk

*Tamias sibiricus* (Rodentia)

**FOUND** *naturally in dense woodland, introduced populations mostly inhabit urban parks.*

Easily recognizable from its combination of five dark brown stripes on the back and face, separated by whitish stripes, on brown fur, the Siberian Chipmunk is a native to the forests of Russia, but is also found in several other areas where they have escaped from captivity. It is an agile climber, although it does most of its foraging on the ground. It builds its nests in burrows or hollow logs, and hoards nuts and seeds to sustain it through the winter.

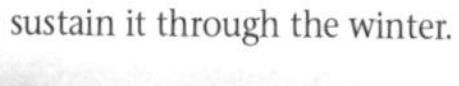

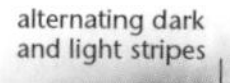

**SIZE** *Body 12–20cm; tail 10–14cm.*
**YOUNG** *One or two litters of 4–6; May–July.*
**DIET** *Nuts, seeds, fruit, fungi, and other plant material; insects and small vertebrates.*
**STATUS** *Common; scarce when introduced.*
**SIMILAR SPECIES** *Barbary Ground-squirrel (p.55), which has a less distinct pattern.*

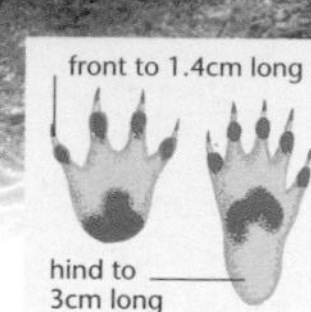


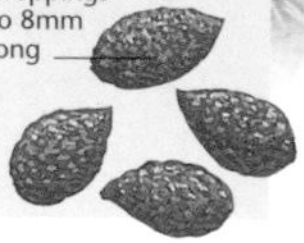

# Lesser Mole-rat

*Nannospalax leucodon* (Rodentia)

**FAVOURS** *dry grassland and cultivated areas; look for "mole-hills" and food plants that are often seen dragged down into connecting burrows.*

Being both nocturnal and exclusively tunnel-dwelling, the Lesser Mole-rat is rarely seen, accounting for some difficulties in deciding just how many species they are and what they should be called. But once seen, the adaptations to its lifestyle are clear: a cylindrical body, with no visible eyes, ears, or tail, and enlarged lower incisors, used for digging the tunnels.

**NOTE**

*The only species in this book not represented by a photograph, an indication of its highly cryptic lifestyle.*

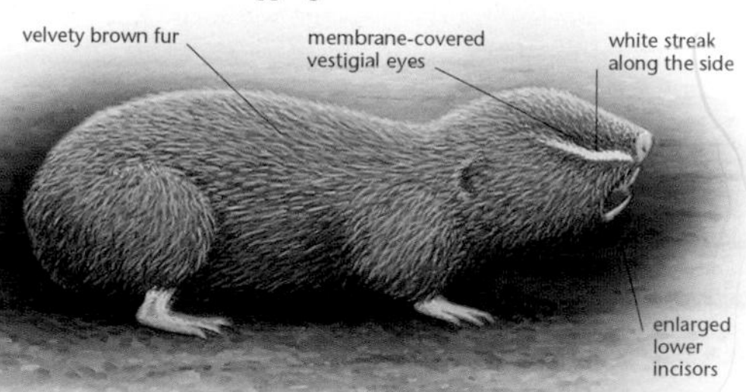


**SIZE** *Body 15–27cm.*
**YOUNG** *Single litter of 2–4; May–June.*
**DIET** *Roots, bulbs, and aerial vegetation.*
**STATUS** *Vulnerable; very locally common.*
**SIMILAR SPECIES** *Moles (pp.27–28), which are insectivores; Balkan Mole-rat* (Spalax graecus)*, which is slightly larger.*

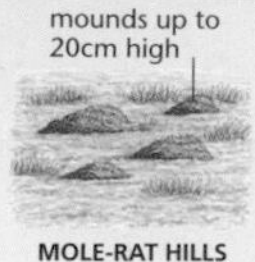

MOLE-RAT HILLS

# Common Hamster

*Cricetus cricetus* (Rodentia)

As a largely nocturnal, burrow-dwelling rodent, the Common Hamster is infrequently seen, despite its occurrence, sometimes in pest proportions, in cultivated fields. It is relatively solitary, living in a complex burrow system, to 2m deep, with separate sleeping areas, latrines, and ladders, which are replenished with food carried in its cheek pouches. Common Hamsters are plump and have a distinctive three-colour pattern: the reddish brown upperparts contrast with largely black underparts, and patches of white fur on the face and flanks.

**BURROWS** *in dry soil without many stones, largely under lowland grassland, steppes, and cultivated fields.*

**NOTE**

*Common Hamsters hibernate between October and March, but they must wake up at approximately weekly intervals to feed on stored food.*

front print to 1.5cm long

hind print to 2cm long

droppings to 1cm

short legs with white feet

**SIZE** *Body 23–24cm; tail 3–5cm.*
**YOUNG** *Up to three litters of 4–12; May–August.*
**DIET** *Seeds, roots, and tubers, and a small proportion of insects.*
**STATUS** *Rare and declining, especially in the west.*
**SIMILAR SPECIES** *Lemmings (p.60), which have much smaller ears; Romanian Hamster* (Mesocricetus newtoni)*, which is smaller, with less black below, and a more easterly range.*

# Norway Lemming

*Lemmus lemmus* (Rodentia)

**FAVOURS** *tundra and alpine zones further south; in years of abundance, spreads into woodland and cultivated areas.*

Norway Lemmings display "boom and bust" population dynamics, typical of many Arctic species; when densities are high and food resources scarce, mass emigrations take place. Active by night and day, the variable black, brown, and yellowish pattern on a rotund, almost tailless body is unmistakable.

front print to 1.2cm long

hind print to 1.8cm long

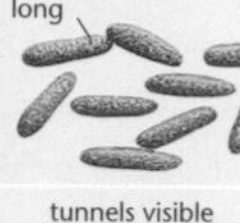


tunnels visible after thaw

TUNNELS

**SIZE** *Body 10–15cm; tail 1–2cm.*
**YOUNG** *Two to six litters of up to 12; April–October.*
**DIET** *Grazes and browses grasses, sedges, dwarf shrubs, and mosses.*
**STATUS** *Common, but fluctuates in density.*
**SIMILAR SPECIES** *Common Hamster (p.59).*

# Wood Lemming

*Myopus schisticola* (Rodentia)

**INHABITS** *moist old-growth coniferous woodland, with extensive ground cover of mosses; drier areas used at times of abundance.*

With its slate-grey coat, and rusty brown rump patch on adults, the Wood Lemming is less striking than the larger Norway Lemming. However, its breeding system is unusual – males make up only a quarter of the population, and some females can give birth only to other females.

**SIZE** *Body 8–11cm; tail 1–2cm.*
**YOUNG** *Litters of up to six.*
**DIET** *Predominantly mosses.*
**STATUS** *Near-threatened; locally common but declining through habitat loss.*
**SIMILAR SPECIES** *Common Hamster (p.59); Grey-sided Vole (p.66), which has a large tail.*

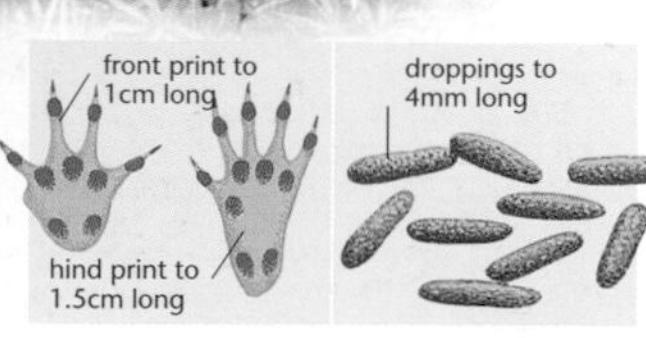

# European Beaver

*Castor fiber* (Rodentia)

The largest of the European rodents, the European Beaver spends all its life in or in close proximity to fresh water. It is supremely adapted to an aquatic lifestyle, with webbed hind feet, a broad scaly tail for steering and propulsion, insulated waterproof fur, and a nose and ears that can be closed while swimming. Beavers are gregarious and are active throughout the year. They are largely nocturnal, spending the day in a lodge or riverside burrow.

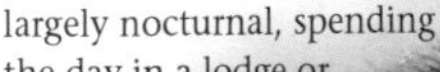

**FREQUENTS** *broad, slow-moving rivers and ponds, usually in flood plains and with trees around the margins.*

small ears

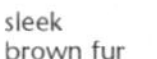

sleek brown fur

### NOTE

*European Beavers are noted for the range of field signs they provide: stick lodges (up to 15m wide), occasionally dams across streams, felled bankside trees, and obvious trackways through vegetation.*

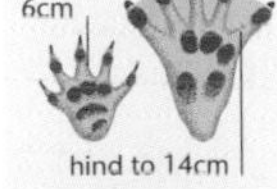

broad, flat, scaly tail

FIELD SIGN

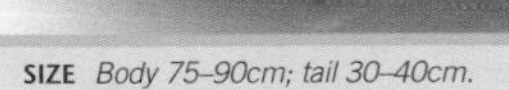

LODGE

**SIZE** *Body 75–90cm; tail 30–40cm.*
**YOUNG** *Single litter of 1–4, June.*
**DIET** *Grasses, wetland vegetation, bark, and wood.*
**STATUS** *Near-threatened; now locally common in parts of its former range as a result of reintroduction schemes.*
**SIMILAR SPECIES** *Coypu, Muskrat, and Water Voles (pp.62–64), which lack the flattened tail; Canadian Beaver (p.62), which is darker.*

# Coypu

*Myocastor coypus* (Rodentia)

**INHABITS** *wetland and waterside habitats, including reedbeds, extending into upper estuarine areas.*

Introduced to Europe from South America, this large aquatic rodent is intermediate in size between Muskrat and Beaver. An effective swimmer, its thick underfur, or nutria, for which the Coypu was first brought to Europe, is waterproof. It has partly webbed hind feet. Mainly nocturnal and crepuscular, it spends its daylight hours in burrows.

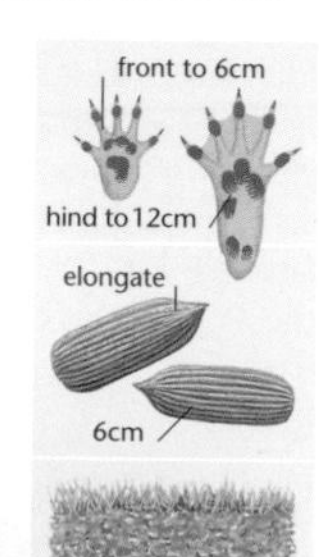

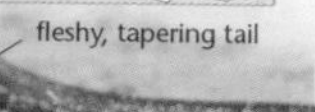


SWIMMING

**SIZE** *Body 35–60cm; tail 25–45cm.*
**YOUNG** *Two or three litters of 4–8.*
**DIET** *Aerial parts and rhizomes of aquatic plants and root crops.*
**STATUS** *Locally common.*
**SIMILAR SPECIES** *European Beaver (p.61), Muskrat (p.63) do not have cylindrical tails.*

# Canadian Beaver

*Castor canadensis* (Rodentia)

**FOUND** *in ponds and small rivers, especially in woodland; greater tendency to block rivers with dams than the European Beaver.*

Similar to its native European counterpart, the Canadian Beaver is usually darker in colour. These beavers have established themselves in several areas through escapes from fur farms; although there is no suggestion of inter-breeding with the native species, it is believed that Canadians outcompete Europeans where they occur together.

**SIZE** *Body 55–85cm; tail 25–30cm.*
**YOUNG** *Single litter of 3–4; June–July.*
**DIET** *Woody and herbaceous plant matter.*
**STATUS** *Introduced from North America; locally common in Finland.*
**SIMILAR SPECIES** *European Beaver (p.61).*

# Muskrat

*Ondatra zibethicus* (Rodentia)

Rather like an over-sized Water Vole, the Muskrat is an aquatic rodent. Its thick, waterproof fur provides insulation, while the vertically flattened tail (three times deeper than wide) and partially webbed hind feet, effectively enlarged with a fringe of stiff bristles, are used for propulsion. When swimming, only the head and shoulders are visible, unlike water voles in which the entire back can be seen.

sleek, brown fur

**OCCURS** *in freshwater habitats; bankside burrows and winter lodges of grass and reeds.*

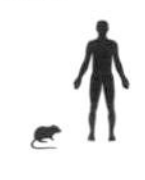

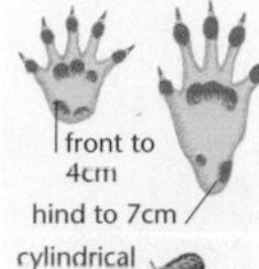

**WINTER LODGE**

**SIZE** *Body 25–40cm; tail 20–30cm.*
**YOUNG** *One to three litters of up to 11.*
**DIET** *Aquatic vegetation, and occasionally molluscs and fish.*
**STATUS** *Common where established.*
**SIMILAR SPECIES** *Coypu (p.62), Water Voles (below and p.64).*

# South-western Water Vole

*Arvicola sapidus* (Rodentia)

The South-western Water Vole is dark grey-brown, with a blunt nose and small ears. In places where it overlaps with the (more northern) Water Vole, it is larger and darker, and, at more than 11cm, generally has a longer tail. However, skull measurements or chromosome counts are needed to differentiate between the two, as some Water Voles are often equally large and equally dark in colour.

**LIVES** *in freshwater habitats, including rivers, canals, lakes, and marshes.*

blunt nose
dark grey-brown fur

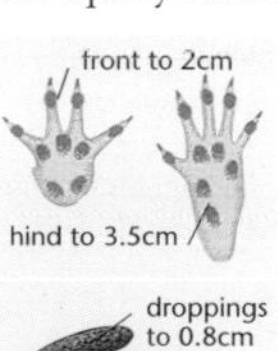


droppings to 0.8cm

**SIZE** *Body 17–20cm; tail 11–13cm.*
**YOUNG** *Two to four litters of up to six.*
**DIET** *Aquatic vegetation and invertebrates, fish, and amphibians.*
**STATUS** *Near threatened; declining.*
**SIMILAR SPECIES** *Muskrat (above); Water Vole (p.64), which has a shorter tail.*

# Water Vole

*Arvicola terrestris* (Rodentia)

**FOUND** *in freshwater habitats – listen for the "plop" as it dives for safety; also in meadows and pastures.*

A large, blunt-nosed vole with a furry tail, the Water Vole is usually found in aquatic habitats, except where it overlaps in range with the South-western Water Vole, where it adopts a more terrestrial existence. Usually seen swimming, its rather frantic doggy-paddle is in marked contrast to the powerful motion of the larger aquatic rodents. The fur is variable in colour, usually brown but in parts of the range a large proportion of voles can be almost black. Shredded bark, cut grass leaves, and irregularly gnawed nuts are often the best evidence of its presence, as it is largely crepuscular.

FEEDING

BURROW

**NOTE**

*Often mistaken for the Common Rat, which also swims well and is found in similar habitats, the Water Vole is also known as the Water Rat. Although a similar size and shape, the Common Rat can be recognized by its pointed snout and much longer tail.*

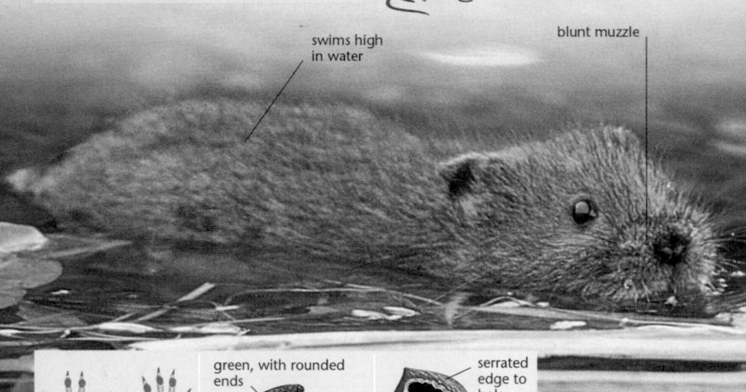


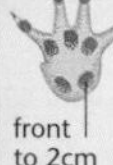

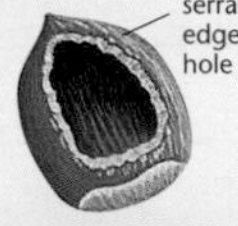


GNAWED NUT

**SIZE** *Body 12–22cm; tail 6–12cm; largest in the north.*
**YOUNG** *Two to five litters of 2–6, March–October.*
**DIET** *Grasses, sedges, and other vegetation, including bulbs, roots, and crops; strips bark in winter.*
**STATUS** *Common, but declining due to pollution and habitat loss.*
**SIMILAR SPECIES** *South-western Water Vole (p.63); where living terrestrially, other voles (pp.65–72), though twice the size of most of these.*

# Bank Vole

*Clethrionomys glareolus* (Rodentia)

The most widespread of the "red-backed voles", this blunt-nosed vole is found in high densities in riverine woodland. It climbs trees well, making it easier to observe than most other, less arboreal voles. However, since much of its activity is under the cover of darkness, an appreciation of its feeding signs is important. These range from small cones stripped of scales and seeds (looking less frayed than those eaten by squirrels) to hazelnuts with a large, neat, round hole, and no teeth marks outside the opening.

**LIVES** *in grassland with plenty of scrub, deciduous, and mixed shrubby woodland.*

large ears and eyes

russet brown back

pale underparts

front to 1cm

hind to 1.8cm

droppings to 4mm

**NOTE**

*There are more than 30 subspecies in Europe, varying in size, colour, skull shape, and teeth; some island sub-species are up to a third larger than mainland ones.*

grey-brown on flanks

tail half head and body length

NIBBLED NUT

inner bark

AERIAL BARK STRIPPING

**SIZE** *Body 8–11cm; tail 3.5–7cm.*
**YOUNG** *Four or five litters of 3–5; April–October, or year round if conditions are favourable.*
**DIET** *Buds, leaves, fruit, seeds, and fungi; some invertebrates.*
**STATUS** *Common; marked population fluctuations in some areas.*
**SIMILAR SPECIES** *Red and Grey-sided Voles (p.66); Field Vole (p.68); other* Microtus *species, which are less red and have shorter tails.*

# Grey-sided Vole

*Clethrionomys rufocanus* (Rodentia)

**INHABITS** *shrub-covered tundra in the north and on mountain tops further south; also open coniferous and birch woodland.*

A northern species, the Grey-sided Vole is distinguished from the related Bank and Red Voles by its reddish fur restricted to a narrow zone down its back, the sides and flanks appearing much greyer. It is also the largest of the three species. As is typical with Arctic species, its population fluctuates markedly, on a cycle of 4–5 years.

reddish band of fur

large eyes

front print to 1cm

hind print to 1.8cm

**SIZE** *Body 11–13cm; tail 3–4cm.*
**YOUNG** *Two or three litters of up to 11.*
**DIET** *Buds and shoots of dwarf shrubs; fruit, berries, and bark; fungi and insects.*
**STATUS** *Common in its restricted range.*
**SIMILAR SPECIES** *Red Vole (below); Bank Vole (p.65); Wood Lemming (p.60).*

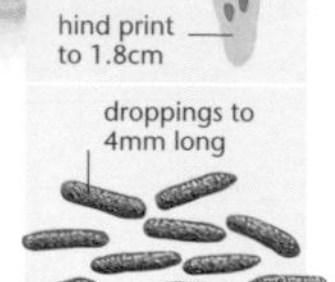

droppings to 4mm long

# Red Vole

*Clethrionomys rutilus* (Rodentia)

**FAVOURS** *sub-Arctic birch and willow woodland, with a well-developed grass and herb layer; mossy coniferous woodland.*

Much brighter than a Bank Vole, the Red Vole also has a shorter tail, which is thick, with a dense covering of hair and a terminal tuft. It is an inhabitant of the northern forest zone, where it seems to avoid felled areas, unlike its more abundant relatives. Nests are built usually under tree roots or stones or in holes in trees. They are lined with grass, and in autumn, are used to store seeds for winter.

red-brown upperparts

short furry tail

front print to 1cm

hind print to 1.7cm

**SIZE** *Body 8–11cm; tail 2.5–3.5cm.*
**YOUNG** *Two or three litters of up to 11.*
**DIET** *Shoots, buds, nuts, and seeds; also bark, lichens, and fungi.*
**STATUS** *Locally common.*
**SIMILAR SPECIES** *Grey-sided Vole (above); Bank Vole (p.65).*

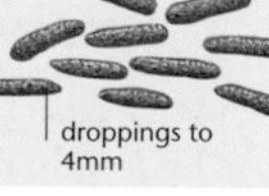

droppings to 4mm

# Snow Vole

*Chionomys nivalis* (Rodentia)

Showing a fragmented distribution largely in the mountains of southern Europe, the Snow Vole is assumed to be a glacial relict species. After the last Ice Age, it was found over wider areas, but has become restricted to cooler mountain tops. It is the highest altitude vole in Europe, remaining active throughout the year, making burrow systems under snow which are then revealed in spring as the snow melts. The Snow Vole is very pale, with a long whitish tail, which appears relatively thick because of its dense covering of hair.

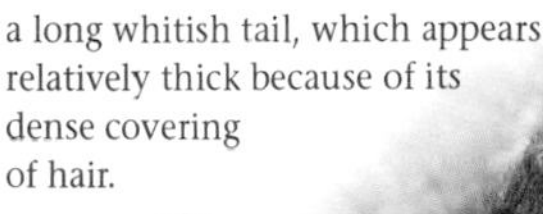

**OCCUPIES** *mountain areas, largely above the tree-line, especially among scree and scrub; in some areas, also found in low, rocky hills, especially on limestone.*

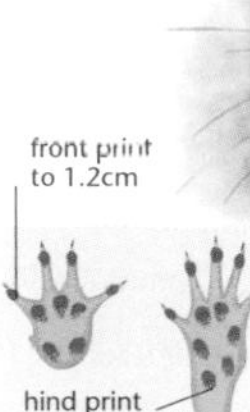


prominent whiskers, to 6cm long

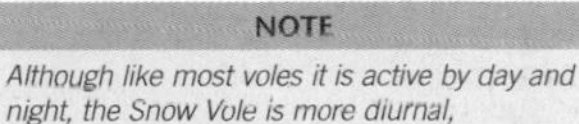

**NOTE**

*Although like most voles it is active by day and night, the Snow Vole is more diurnal, particularly during the summer, and is often to be seen sunbathing on rocks in spring.*

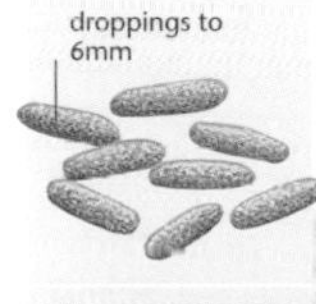


pale grey-brown fur

**BURROW TRAILS AFTER THAW**

pale, long, hairy tail

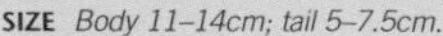

**SIZE** *Body 11–14cm; tail 5–7.5cm.*
**YOUNG** *One or two litters of up to seven, May–September.*
**DIET** *Grass and other vegetation, including the shoots and fruit of dwarf shrubs.*
**STATUS** *Near-threatened; locally common.*
**SIMILAR SPECIES** *Balkan Snow Vole* (Dinaromys bogdanovi), *which is paler grey, with larger ears, and an even longer but thinner tail.*

# Common Vole

*Microtus arvalis* (Rodentia)

**THRIVES** *in habitats with reduced cover, such as dry, short, or grazed grassland, and cultivated fields.*

More of a burrowing species than the Field Vole, the Common Vole is largely nocturnal, so look out for its burrows (with spoil heaps), feeding piles, and droppings. It has a short tail which is marginally darker above than it is below. In some instances, identification rests on a detailed examination of its teeth.

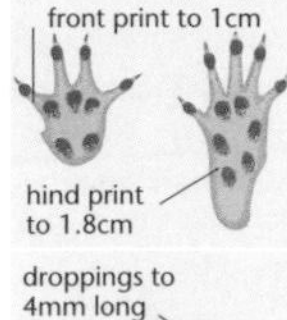

BURROW

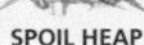

SPOIL HEAP

**SIZE** *Body 9–12cm; tail 3–4cm.*
**YOUNG** *Two to four litters of up to 12.*
**DIET** *Grass and roots, especially of rushes.*
**STATUS** *Common.*
**SIMILAR SPECIES** *Field Vole (below); Root and Cabrera's (p.69) Voles; Sibling Vole* (M. rossiaemeridionalis), *is its eastern counterpart.*

# Field Vole

*Microtus agrestis* (Rodentia)

**FOUND** *in marshes, tall grassland, and open woodland, shreds leaves and grass forming clear pathways.*

Also known as the Short-tailed Vole, the Field Vole has a tail that is shorter than many other voles, and is distinctly dark above. Its long, shaggy fur almost covers its small ears, which are partially hairy inside, especially at the base. Differentiation of the *Microtus* voles by their characteristic teeth details, is of little value in the field, but useful for remains found in owl pellets.

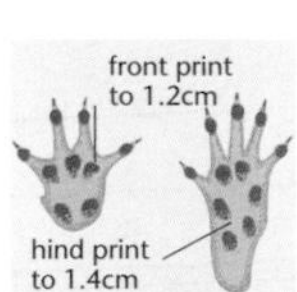


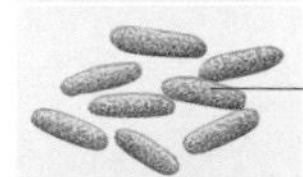

NEST

dark grey-brown fur

short, rounded muzzle

**SIZE** *Body 8–13cm; tail 2–5cm.*
**YOUNG** *Three to seven litters of 2–6 young.*
**DIET** *Grass, roots, fungi, and bark.*
**STATUS** *Common; may be a pest.*
**SIMILAR SPECIES** *Common Vole (above); Bank Vole (p.65); Root and Cabrera's (p.69) Voles.*

# Root Vole

*Microtus oeconomus* (Rodentia)

Larger than the Field Vole, the Root Vole has darker brown fur. Its tail, longer than the Field Vole, is typically less than half of the head/body length, distinctly dark above, and pale below. Its ears are less hairy inside than those of Field Vole. It is found in damper situations than other *Microtus* species, and both swims and climbs well. It's grass nests are usually made above the ground.

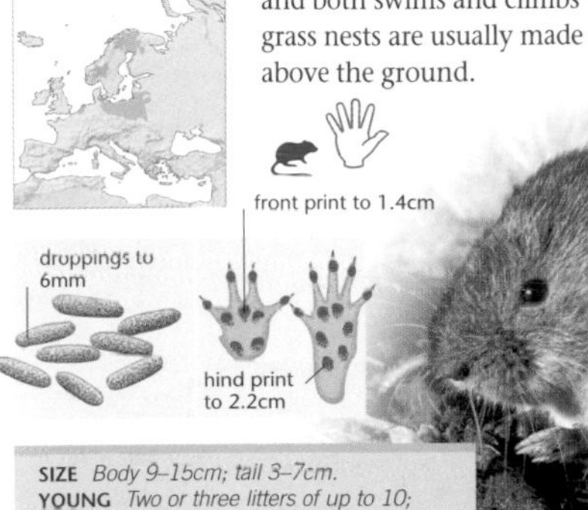


**LIVES** *in moist areas, damp grassland, and sedge beds; occupies drier grassland and sometimes, buildings in winter.*

ears covered with hair

dark brown fur

**SIZE** *Body 9–15cm; tail 3–7cm.*
**YOUNG** *Two or three litters of up to 10; May–September.*
**DIET** *Shoots, leaves, and buds.*
**STATUS** *Near-threatened; locally common.*
**SIMILAR SPECIES** *Cabrera's Vole (below); Common and Field Voles (p.68).*

# Cabrera's Vole

*Microtus cabrerae* (Rodentia)

Very similar to the Common Vole, Cabrera's Vole is a large species. Where the two species overlap, Common tends to be found at higher altitudes than Cabrera's. A slightly darker coat, with long, dark hairs protruding from the hindquarters, and a buff suffusion to the underparts are reliable indicators of Cabrera's Vole, but teeth and skull features are needed to confirm identification.

**FAVOURS** *lowland grassy and marshland habitats, as well as open woodland, typically with a dense shrub layer.*

dark brown upper fur

grey-buff underparts

front print to 1.4cm

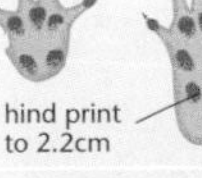

hind print to 2.2cm

droppings to 4mm

**SIZE** *Body 11–13cm; tail 3–5cm.*
**YOUNG** *Poorly known; litters of 3–5.*
**DIET** *Grass and other vegetation.*
**STATUS** *Near-threatened; scarce.*
**SIMILAR SPECIES** *Root Vole (above); Field and Common Voles (p.68); Guenther's Vole* (M. guentheri), *which has a pale tail and feet.*

# Common Pine Vole

*Microtus subterraneus* (Rodentia)

**OCCURS** *in a wide range of grassland and woodland habitats, both wet and dry, lowland and upland.*

**NOTE**

*Pine voles differ from grass voles in their smaller eyes, ears, and feet (a reflection of their more subterranean habits). Pine voles are thus sometimes classed in a separate genus –* Pitymys.

The most widespread of the pine voles, the Common Pine Vole is found in all kinds of grassland and woodland habitat across much of Europe. Its shape and diet reflect its largely burrowing habits, which along with its predominantly nocturnal activity make it easy to overlook. Nesting and food storage takes place within the tunnel systems, which are sometimes sealed up against inclement wet or cold weather conditions. Pine voles have five pads on their hind feet and two pairs of nipples unlike grass voles within the same genus *Microtus,* which have six pads on their hind feet and four pairs of nipples.

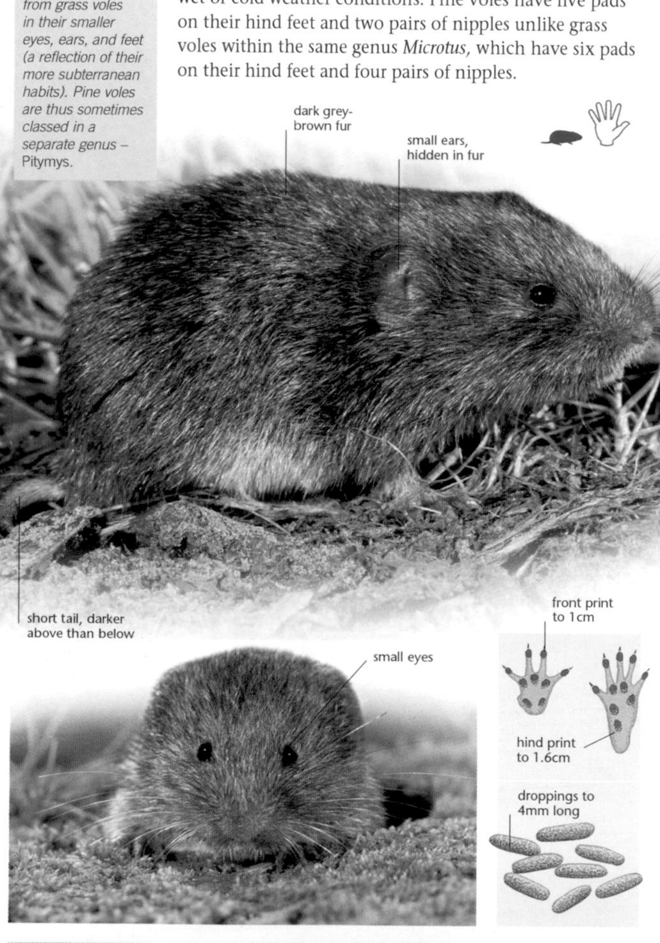

**SIZE** *Body 8–10.5cm; tail 2.5–4cm.*
**YOUNG** *Up to 10 litters of 2–4, all year round.*
**DIET** *Largely the underground parts of plants.*
**STATUS** *Common.*
**SIMILAR SPECIES** *Other pine voles (pp.71–72); Tatra Pine Vole* (M. tatricus), *which is larger and darker, but with a very restricted range in the mountains of E. Europe.*

# Mediterranean Pine Vole

*Microtus duodecimcostatus* (Rodentia)

A strongly tunnel-dwelling, largely nocturnal vole, the Mediterranean Pine Vole is widespread across the Iberian Peninsula and southern France. It is found primarily where the soil is loose and deep enough for tunnelling, and given its distribution, it is more tolerant of drought and high temperatures than most other voles. Its yellowish coat with a velvety texture is a helpful identification feature on a dead or trapped specimen.

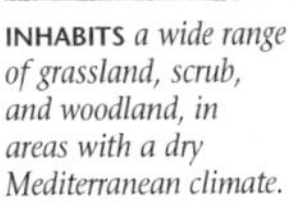

**INHABITS** *a wide range of grassland, scrub, and woodland, in areas with a dry Mediterranean climate.*

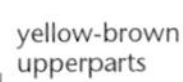


dense, short, velvety fur

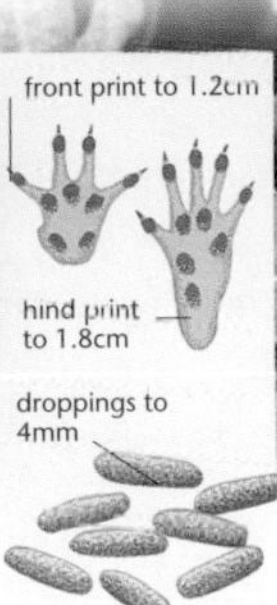


droppings to 4mm

**SIZE** *Body 8.5–10.5cm; tail 2–3cm.*
**YOUNG** *Poorly known; litter size up to five.*
**DIET** *Grasses, other plants, and crops.*
**STATUS** *Common.*
**SIMILAR SPECIES** *Pine voles (pp.70–72); Lusitanian Pine Vole* (M. lusitanicus) *and Thomas's Pine Vole* (M. thomasi).

# Savi's Pine Vole

*Microtus savii* (Rodentia)

Very similar to the Common Pine Vole, Savi's Pine Vole can be distinguished by the simple nature of the ridges on its outer molars. It has a paler brown coat, shorter tail, and even smaller ears. It is found throughout the Italian peninsula, where other pine voles are absent, and can reach very high densities.

**FOUND** *in a wide range of grassland, scrub, open woodland, and montane habitats, as well as cultivated and urban areas.*

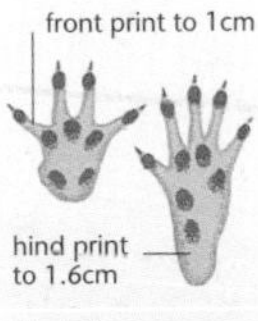


droppings to 4mm

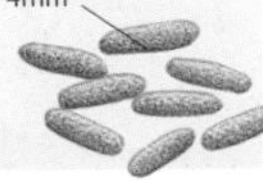

**SIZE** *Body 7.5–10.5cm; tail 2–3.5cm.*
**YOUNG** *Poorly known; litter size of 2–4.*
**DIET** *All plant parts, underground and aerial.*
**STATUS** *Common.*
**SIMILAR SPECIES** *Other pine voles (pp.70–72); Balkan Pine Vole* (M. feltenii), *which differs in range and chromosomes.*

# Pyrenean Pine Vole

*Microtus gerbei* (Rodentia)

**TUNNELS** *in grassland and cultivated areas, extending into montane scrub and open woodland; especially in lower rainfall areas.*

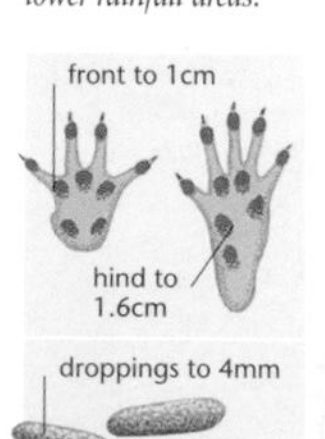


droppings to 4mm

Overlapping in distribution with several other pine voles, and distinguishable from Savi's only by its chromosomes, the Pyrenean Pine Vole is very hard to identify due to its extreme burrowing habit, especially in lowlands. It has a short tail and its coat is intermediate in colour between the darker Common and paler Mediterranean Pine Voles.

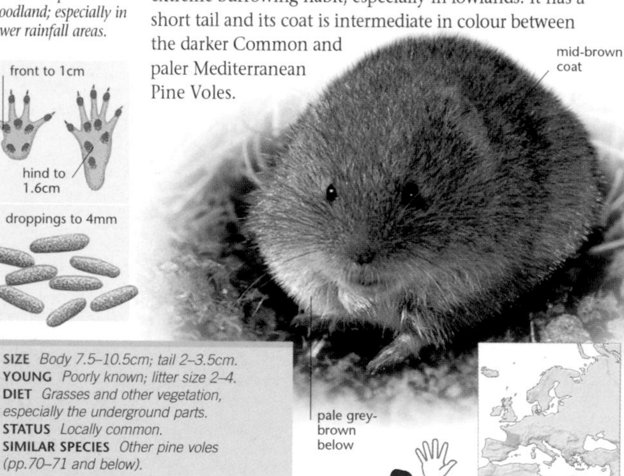


**SIZE** *Body 7.5–10.5cm; tail 2–3.5cm.*
**YOUNG** *Poorly known; litter size 2–4.*
**DIET** *Grasses and other vegetation, especially the underground parts.*
**STATUS** *Locally common.*
**SIMILAR SPECIES** *Other pine voles (pp.70–71 and below).*

# Alpine Pine Vole

*Microtus multiplex* (Rodentia)

**FAVOURS** *open woodland, especially in mountainous areas; also dry meadows and cultivated areas in lowland.*

Externally very similar to other pine voles in Europe, the Alpine Pine Vole is generally a little larger and displays a wider variety of yellowish and reddish tones in its fur. Definitive identification is, however, outside the sphere of the field naturalist because it involves chromosome counts and similar studies. Despite its name, this vole is not restricted to mountains, occurring down to sea level on the French south coast.

mid-brown coat

short tail

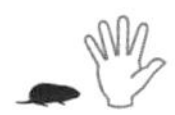

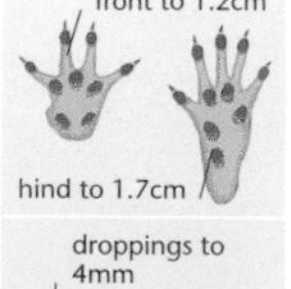


**SIZE** *Body 9–11.5cm; tail 2.5–4cm.*
**YOUNG** *Little known; litter size 2–4.*
**DIET** *Vegetation, especially roots, bulbs, and other underground plant parts.*
**STATUS** *Locally common.*
**SIMILAR SPECIES** *Other pine voles (pp.70–71 and above).*

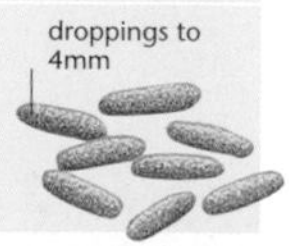

# Hazel Dormouse

*Muscardinus avellanarius* (Rodentia)

A small, orange-brown, largely arboreal rodent, the Hazel or Common Dormouse is nocturnal and hibernates from October to April in a tightly curled position to conserve heat. It is rather sensitive to cold, and even during its active period, cold weather can induce periods of torpor. The nest is a spherical ball of woven grass, moss, and bark strips, positioned in dense undergrowth, or in tree forks or nesting boxes. As befits its climbing habit, its feet are prehensile for grasping branches, and have well-developed pads to enable it to grip well. Its long furry tail may occasionally have a white tip.

**FOUND** *in deciduous, often coppiced, woodland, scrub, and thick hedges; rather elusive, so best located by looking for feeding signs, especially hazelnuts with a neat round hole and a smooth inner margin.*

SLEEPING

**NOTE**

*A classic habitat for the Hazel Dormouse is coppiced hazel, abundant bramble, and climbing honeysuckle, the latter providing bark strips to line its nest with.*

NIBBLED HAZELNUT

**SIZE** *Body 6–9cm; tail 5.5–8cm.*
**YOUNG** *One or two litters of 4–7, June–August.*
**DIET** *Flowers, insects, and fruit that is seasonally available in its habitat.*
**STATUS** *Near-threatened; scarce and declining due to habitat fragmentation.*
**SIMILAR SPECIES** *Harvest Mouse (p.83), which is a similar size and colour, but has a hairless tail.*

# Fat Dormouse

*Glis glis* (Rodentia)

**FOUND IN** *mature deciduous woodland, often of beech or sweet chestnut; does not require a shrub layer.*

Also known as the Edible Dormouse, the Fat Dormouse puts on a thick layer of fat by eating nutritionally rich seeds and nuts in preparation for hibernation. For this reason, it has long been hunted and kept in captivity as a source of food and fur. Its large size and grey fur can lead to its being confused for a squirrel, especially the similarly sized Flying Squirrel, although the overlap in range is limited. It has large eyes, their size exaggerated by the circles of black fur that surround them. This dormouse is nocturnal.

moderately prominent ears

large eyes

grey fur, often tinged brown

long, bushy tail

round eyes

prehensile feet

front to 1.5cm

hind to 3cm

**NOTE**

*Search for nest sites for a glimpse of this dormouse: look for spherical nests of grass, leaves, and moss, often wedged in the fork of a branch or in holes in trees, buildings, or underground.*

**SIZE** *Body 13–19cm; tail 12–15cm.*
**YOUNG** *Single litter of 2–9; June–August.*
**DIET** *Nuts, seeds, fungi, bark, insects, and birds' eggs and nestlings.*
**STATUS** *Near-threatened; locally common, especially in the south.*
**SIMILAR SPECIES** *Grey Squirrel (p.54); Flying Squirrel (p.55).*

# Garden Dormouse

*Eliomys quercinus* (Rodentia)

Although it is an agile climber, the nocturnal Garden Dormouse is rather less strictly arboreal than the other dormice and will nest in holes in rocks and walls. Its reddish brown upperparts contrast with white underparts, and it has a distinctive black face mask. The long, furry tail is tipped black and white, and the ears are particularly prominent.

**INHABITS** *deciduous and coniferous woodland, scrubby and rocky places, and orchards and gardens; often enters houses.*

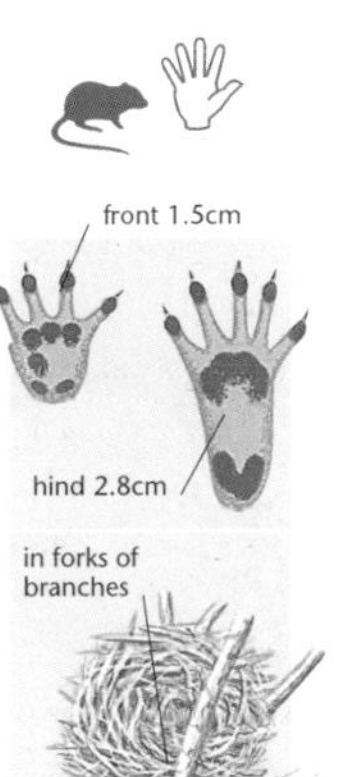


**SIZE** *Body 10–17cm; tail 9–15cm.*
**YOUNG** *One or two litters of 4–6; May–August.*
**DIET** *Largely vegetarian: buds, shoots, fruit, and nuts; occasionally invertebrates.*
**STATUS** *Vulnerable; locally common.*
**SIMILAR SPECIES** *Forest Dormouse (below).*

# Forest Dormouse

*Dryomys nitedula* (Rodentia)

Somewhat similar to the Garden Dormouse, the Forest Dormouse is an altogether more easterly species. Both species have a black face mask, but in the Forest Dormouse it does not extend behind the ears. Its long tail is very bushy but unpatterned. This dormouse nests in holes in trees, rocks, and walls.

**LIVES** *in deciduous and coniferous woodland with a dense shrub layer, cultivated areas, rocky meadows, and gardens.*

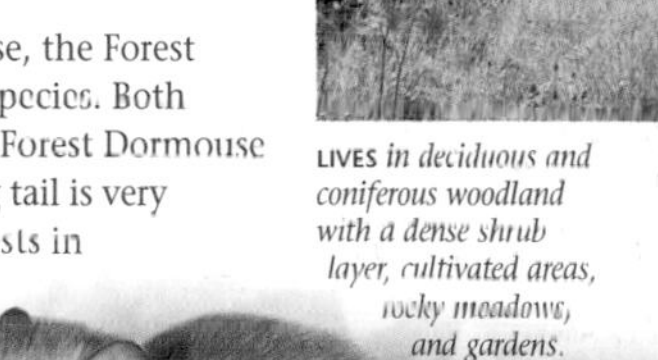

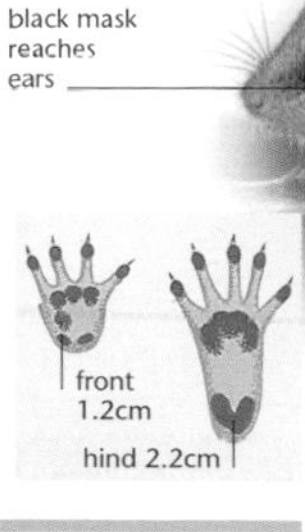


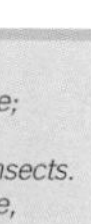

**SIZE** *Body 8–13cm; tail 8–9cm.*
**YOUNG** *One to three litters of up to five; June–July.*
**DIET** *Buds, seeds, fruit, lichens, and insects.*
**STATUS** *Near-threatened; generally rare, though locally common in the Balkans.*
**SIMILAR SPECIES** *Garden Dormouse (above).*

# Northern Birch Mouse

*Sicista betulina* (Rodentia)

**OCCUPIES** *woodland, both deciduous and coniferous, with a dense shrub layer, but especially birch in the north; more montane in the south.*

Birch mice are very long-tailed, small rodents, which display a characteristic bounding gait on the ground, the tail often held curved upwards. The Northern Birch Mouse climbs very well, using its opposable outer toes and semi-prehensile tail for support and balance. A nocturnal animal, hibernation takes place between October and May, in a grass-lined nest within a hollow log, under a tree stump, or sometimes above the ground in dense undergrowth.

**NOTE**

*Relative to its body, the Northern Birch Mouse's tail is longer than any other European rodent; it can be curled round a twig for support while climbing.*

**SIZE** *Body 5–7cm; tail 8–11cm.*
**YOUNG** *One or two litters of up to eight; May–June.*
**DIET** *Insects; some grain and fruit prior to hibernation.*
**STATUS** *Near-threatened; generally rare, but very locally common.*
**SIMILAR SPECIES** *Striped Field Mouse (p.77), which is much larger; Southern Birch Mouse (S. subtilis), which has a shorter tail, a dark stripe bordered with pale fur, and occupies more open habitats.*

# Striped Field Mouse

*Apodemus agrarius* (Rodentia)

Among the true mice, the Striped Field Mouse is unique for the dark stripe running down its back, from the top of the head to the rump. Otherwise, it is rather similar to the Wood Mouse, although with less prominent ears, and with no chest spot on its white underparts. Its tail has 120–140 rings and it has relatively short whiskers. Southern populations tend to be larger than more northerly populations.

**FAVOURS** *lowland regions, occurring on the edges of woodland, in scrub, hedgerows, gardens, and cultivated land; nests underground and some spend the winter in buildings.*

**SIZE** *Body 7.5–12cm; tail 7–8.5cm.*
**YOUNG** *Up to six litters of 5–6; all year.*
**DIET** *Plant material, including buds, fruit, and grain; insects, worms, and molluscs.*
**STATUS** *Scarce to common.*
**SIMILAR SPECIES** *Northern Birch Mouse (p.76); Rock Mouse* (A. mystacinus).

# Wood Mouse

*Apodemus sylvaticus* (Rodentia)

**OCCURS** *in woodland, forest edges (Beech and Spruce), grassland, marshes, rocky areas, and cultivated land; frequently to be found in buildings.*

One of the commonest and most widespread of European mammals, the Wood Mouse is found throughout Britain and continental Europe. Although it is generally assumed that a mouse in a house is a House Mouse, very often it is a Wood Mouse, with larger ears, eyes, and feet to prove it. It is very agile, with a bounding, kangaroo-like gait across open ground, and climbs nimbly up trees. It sometimes has a yellowish chest spot, although if present at all, it is never as extensive as that of the Yellow-necked Mouse, and always longer than it is broad. Feeding signs to look out for are fir cones with scales neatly gnawed off, and hazelnuts with teeth marks around the outside of the hole.

JUVENILE

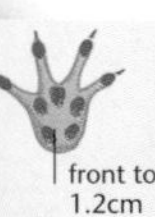

NIBBLED CONE

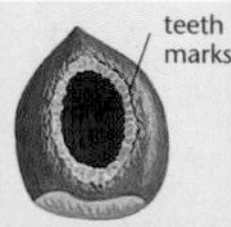

NIBBLED NUT

**NOTE**

*Although largely a nocturnal species, the Wood Mouse is insensitive to infra-red light, and can be clearly observed in filtered torchlight.*

**SIZE** *Body 8–11cm; tail 7–11.5cm.*
**YOUNG** *One or two litters of 4–8; March–October.*
**DIET** *Seeds (especially acorns, beech mast, hazelnuts), cereal crops, buds, and saplings; insects and snails.*
**STATUS** *Common.*
**SIMILAR SPECIES** *Yellow-necked Mouse (p.79); Pygmy Field Mouse (A. uralensis), which is smaller, greyer, and restricted to E. Europe.*

# Yellow-necked Mouse

*Apodemus flavicollis* (Rodentia)

Significantly more robust than a Wood Mouse, the Yellow-necked Mouse is also a richer brown colour above and the brown is divided sharply from the pure white underparts. Its distinguishing feature is a yellow spot or collar on the chest. The amount of yellow on the neck is variable, but always more extensive than in the Wood Mouse. It is also a more arboreal species, often feeding nocturnally in the canopy of deciduous trees. It carries seeds and fruit to underground storage places, which help it to sustain itself and remain at least partially active during the winter months.

**FOUND** *in deciduous woodland, hedges, orchards, and gardens, but largely montane in the south; readily enters houses.*

**NOTE**

*The yellow chest mark, for which this rodent is named, is very variable; sometimes it forms a collar but if it is a spot, it is always more broad than it is long.*

large ears

bulbous eyes

rich brown upperparts

white below

front to 1.5cm

hind to 2.5cm

droppings to 6mm

long tail, with 165–235 rings

**SIZE** *Body 9–13cm; tail 9–13cm.*
**YOUNG** *Three or four litters of up to eight; April–October.*
**DIET** *Seeds, seedlings, buds, fruit, fungi, and invertebrates.*
**STATUS** *Common.*
**SIMILAR SPECIES** *Wood Mouse (p.78); Alpine Mouse* (A. alpicola), *which has chest markings intermediate between those of Yellow-necked and Wood Mice, different skull features, and a relatively longer tail.*

# House Mouse

*Mus domesticus* (Rodentia)

**FOUND** *mainly in buildings, where it is active by night, usually making a bulky nest of shredded paper and fabric, leaving a distinctive musty smell; can occupy a wide range of semi-natural habitats.*

Worldwide the most widely distributed mammal, apart from humans, the House Mouse is variable in colour, typically dark grey-brown in the north and west, and a paler reddish brown further south. The underside is equally variable, from a dusky grey to pale brown. There is no sharp line between the colour of the upper- and underparts. It has relatively small eyes and ears. Its hairless, thick tail has 140–175 prominent rings and is as long as the head and body length. A distinctive feature of its teeth is that the upper incisors have a notch at the tip.

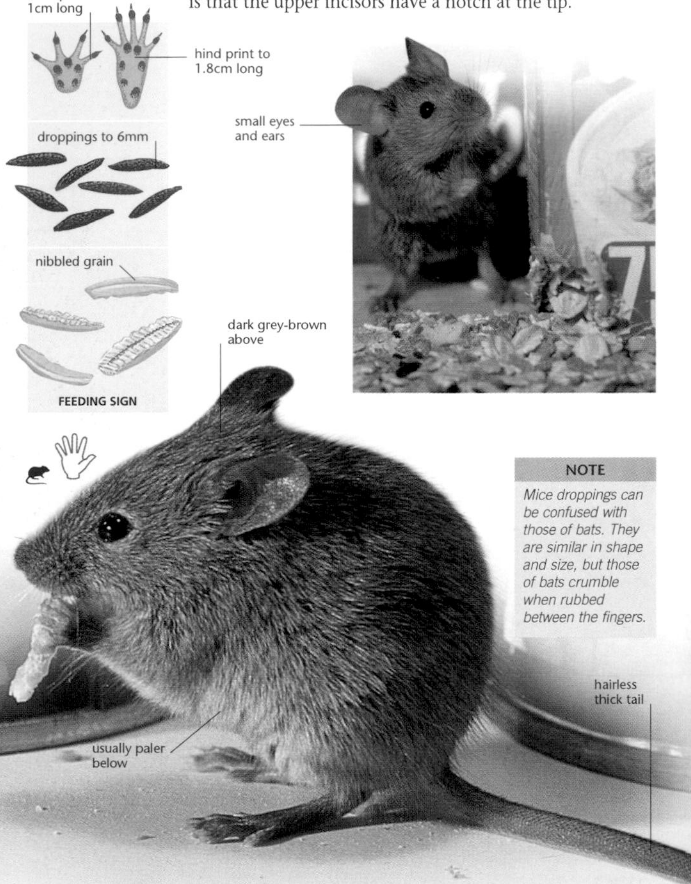


**NOTE**

*Mice droppings can be confused with those of bats. They are similar in shape and size, but those of bats crumble when rubbed between the fingers.*

**SIZE** *Body 7–10cm; tail 7–9.5cm.*
**YOUNG** *Up to 10 litters of 4–8; year-round if sufficient food is available.*
**DIET** *Grain, stored food, and other plant matter; some invertebrates.*
**STATUS** *Common.*
**SIMILAR SPECIES** *Algerian Mouse (p.81); Eastern House Mouse* (M. musculus), *which has up to 200 tail rings; Balkan Short-tailed Mouse* (M. macedonicus), *which is restricted to S.E. Europe.*

# Algerian Mouse

*Mus spretus* (Rodentia)

Unlike the related House Mouse, the Algerian Mouse is an outdoor species, further distinguished by its shorter tail (less than the head-body length) and a clear line of demarcation between its whitish belly and yellow-brown upperparts. The upper incisors do not have the distinct notch characteristic of the House Mouse.

**FREQUENTS** *scrub, open woodland, gardens, and cultivated areas, avoiding buildings.*

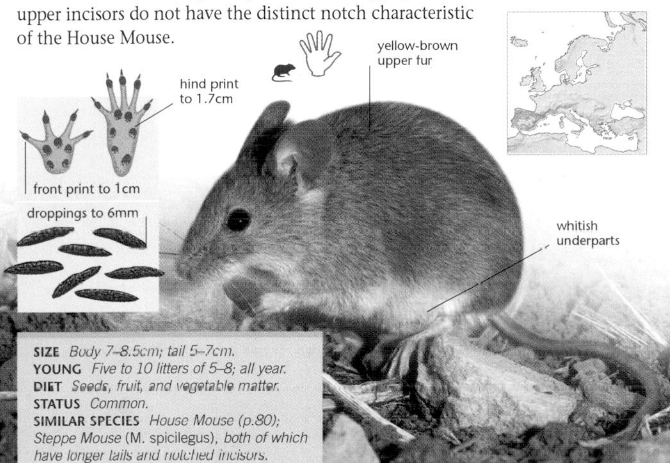


**SIZE** *Body 7–8.5cm; tail 5–7cm.*
**YOUNG** *Five to 10 litters of 5–8; all year.*
**DIET** *Seeds, fruit, and vegetable matter.*
**STATUS** *Common.*
**SIMILAR SPECIES** *House Mouse (p.80); Steppe Mouse* (M. spicilegus), *both of which have longer tails and notched incisors.*

# Ship Rat

*Rattus rattus* (Rodentia)

Also known as the Black Rat, as some forms have very dark fur, the Ship Rat is variable in colour, often appearing grey or brown, and rather shaggy as the coat contains long guard hairs. It is more agile than the Common Rat, readily climbing inside buildings into the roof timbers.

**LIVES** *in and around buildings, especially around sea-ports in northern Europe; the species most frequently found on ships.*

**SIZE** *Body 18–24cm; tail 15–25cm.*
**YOUNG** *Three to five litters of up to 16; March–November.*
**DIET** *Grain; birds' eggs and nestlings.*
**STATUS** *Introduced from Asia; common in the south, but declining elsewhere.*
**SIMILAR SPECIES** *Common Rat (p.82).*

# Common Rat

*Rattus norvegicus* (Rodentia)

**OCCURS** *in urban, industrial, and other developed areas, farms, refuse tips, sewage systems, and around natural river banks.*

The Common Rat is also known as the Brown Rat. Its fur is generally a mid-brown colour, but there is some variability, although less than that shown by the Ship Rat. A robust species, the Common Rat is further distinguished by its thicker tail, which is shorter than the head and body length, as well as being dark above and pale below, and its smaller, brownish, and hairy ears. Largely nocturnal, Common Rats are good swimmers and may be mistaken for Water Voles, although the smaller size and blunt muzzle of the voles are usually obvious. Small juvenile rats can be separated from mice by their disproportionately large feet.

**NOTE**

*Common and Ship Rats can be identified by their droppings: dividing the diameter by the length gives a value of more than 0.4 for Common Rat and less than 0.4 for Ship Rat.*

fleshy tail

pointed muzzle

mid-brown fur, lacking shaggy guard hairs

relatively small hairy ears

front print to 1.8cm

hind print to 3cm

droppings to 2cm

**NIBBLED GRAIN**

**SIZE** *Body 21–29cm; tail 17–23cm.*
**YOUNG** *Up to five litters of up to 15; year round except in cold conditions.*
**DIET** *Seeds, grain, and other vegetable matter, invertebrates and small vertebrates; scavenges from human waste.*
**STATUS** *Introduced from Asia; common.*
**SIMILAR SPECIES** *Ship Rat (p.81); Water voles (pp.63–64).*

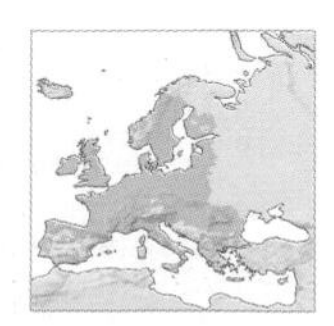

# Harvest Mouse

*Micromys minutus* (Rodentia)

The smallest European rodents, Harvest Mice are very agile climbers, using their feet with opposable outer toes to grip grass stems, and the semi-prehensile tail for additional support. The end 2cm of the tail can be curled around a stem, or held out straight for balance. Mainly nocturnal, they remain active throughout the year, although the winter months are spent largely on or below the ground. The soles of the feet of a Harvest Mouse are sensitive to vibration, and can thus alert them to oncoming danger.

**FOUND** *on long, coarse grassland, scrub patches, river banks, drier edges of reedbeds, and among cereal crops.*

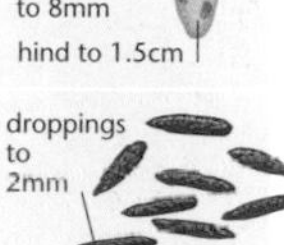

NOTE

*The Harvest Mouse makes a distinctive nest of grass, woven into a ball up to 10cm wide and lodged in vegetation usually 30–60cm above the ground; given their small size and secretive nature, searching for nests in winter is the best way of establishing their presence on a site.*

short hairy ears

whitish underparts

orange-brown fur above

long, semi-prehensile tail

**SIZE** *Body 5–8cm; tail 5–7cm.*
**YOUNG** *Three to seven litters of up to eight; May–October.*
**DIET** *Grain and other seeds; buds, flowers, and fruit; insects and occasionally young birds and rodents.*
**STATUS** *Near-threatened; locally common but declining.*
**SIMILAR SPECIES** *Hazel Dormouse* *(p.73)**, which is only a little larger, but has a bushy tail, and is generally found in wooded areas.*

**DIGS** *deep burrows in dry open woodland, farmland, and Mediterranean scrub habitats; sometimes lives in caves or takes over Badger setts.*

# Crested Porcupine

*Hystrix cristata* (Rodentia)

The Crested Porcupine is an unmistakable large rodent, covered in black and white quills, up to 40cm long. These are used offensively, the porcupine reversing towards a threat with spines raised, and defensively, when it rolls into a ball if the threat persists. Although the spines are not barbed as in some American porcupines, they can inflict a nasty wound; they are easily detached, and can often be found around the entrances to the burrows. The Crested Porcupine is the only European representative of an African and Asian family. It is suggested that the current European population may be derived from Roman introductions from North Africa.

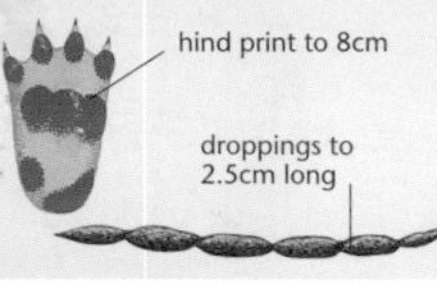


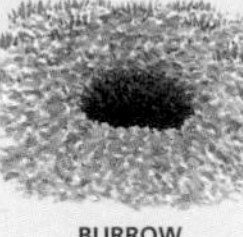


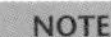

**SIZE** *Body 60–90cm; tail 8–16cm.*
**YOUNG** *One or two litters of up to four; mostly in the summer.*
**DIET** *Roots, bulbs, fruit, and bark (they can be pests in gardens and orchards); carrion.*
**STATUS** *Possibly introduced from North Africa; rare but slowly increasing its range.*
**SIMILAR SPECIES** *None.*

**NOTE**

*The Crested Porcupine can be distinguished from related species that escape from captivity by the long whitish bristles, which run down the back of its head.*

# Brown Bear

*Ursus arctos* (Carnivora)

Hunted to extinction over much of its former European range, the Brown Bear is a large, heavily built mammal with shaggy brown fur and a short tail. The muzzle is elongated with sharp teeth, as befits its occasional carnivorous habits, although much of its diet is vegetarian. The large feet with powerful claws are put to good use in digging out bees' nests, to gain access to one of its favourite foods – honey. This bear is mainly nocturnal, especially where persecuted. It spends winter in a cave or rock crevice, after fattening on berries and nuts, although it does not enter true hibernation.

**FREQUENTS** *extensive, often rocky, forests, tundra and mountain pastures; sometimes forages in agricultural areas and increasingly in refuse bins near the edge of settlements.*

**CUB**

very short tail, hidden in fur

projecting muzzle

shaggy brown fur

**NOTE**

*Given the Brown Bear's solitary and secretive nature, field signs are the best way of establishing its presence: huge footprints, large, rounded droppings, scratch marks on tree trunks, and excavated bees' nests.*

long claws

front print broader, to 25cm long

hind print to 30cm long

droppings to 12cm long

**SIZE** *Body 1.7–2.4m; tail 6–20cm.*
**YOUNG** *Single litter of up to three; January–February.*
**DIET** *Omnivorous; bulbs, fruit, fungi, fish, eggs, honey, and carrion; occasionally predatory, including on domestic stock.*
**STATUS** *Rare/extinct in W. Europe; locally common in N. and E. Europe.*
**SIMILAR SPECIES** *A young bear may be mistaken for a Wolverine (p.90), which has an obvious tail.*

# Polar Bear

*Ursus maritimus* (Carnivora)

An unmistakable large white bear, the Polar Bear has a thick layer of insulating fat that enables it to withstand cold temperatures and help it to survive many hours of swimming between ice floes. Other adaptations to its extreme environment include small ears (to reduce heat loss), black skin (to absorb heat from the sun), and furry soles to its feet.

**INHABITS** *the high Arctic coastline, wandering extensively over sea-ice, usually with easy access to open water.*

**SIZE** *Body 1.6–2.5m; tail 8–10cm.*
**YOUNG** *Single litter of 2 or 3; born winter.*
**DIET** *Seals, fish, Mountain Hares, young or weak Musk-ox and Reindeer; carrion.*
**STATUS** *Conservation-dependent; scarce but stable.*
**SIMILAR SPECIES** *None.*

# Golden Jackal

*Canis aureus* (Carnivora)

Similar in shape to a dog, the Golden Jackal is intermediate in size between the Wolf and the Red Fox. The sandy grey coat, often tinged red, provides effective camouflage in its arid habitats. Largely nocturnal, jackals spend the day hiding in thick vegetation or a burrow; although relatively solitary, they do howl communally, but with a clearer and less mellow tone than wolves.

**ROAMS** *over semi-arid, open grassland, cultivated areas, and marshes.*

**SIZE** *Body 65–100cm; tail 20–30cm.*
**YOUNG** *Single litter of 4–6; May–June.*
**DIET** *Rodents, birds, fruit, insects, and carrion, including scraps from human refuse.*
**STATUS** *Locally common.*
**SIMILAR SPECIES** *Wolf (p.87), which is larger; Red Fox (p.89), which is smaller.*

# Wolf

*Canis lupus* (Carnivora)

Although still widespread across eastern Europe, the Wolf has been driven to extinction in vast parts of its former range in the west, largely the victim of persecution for perceived harm to stock. In these more enlightened times, various reintroduction projects are underway to restore some of that range. Wolves are social animals, living in family groups, which hunt co-operatively in order to tackle large prey, and communal, mournful howling is frequently heard, especially at night. Closely related to domestic dogs, hybridization poses a risk to Wolf populations in some areas.

**NOTE**

*Though similar in size and colour to German Shepherd dogs, Wolves hold their head in line with the haunches, and have erect ears and drooping tails that never curl.*

**OCCUPIES** *open woodland, both deciduous and coniferous, mountain and tundra; increasingly moving near to human habitation, especially when food is scarce.*

front to 8cm

hind to 7cm

tapering droppings to 12cm

short, blunt, upright ears

CUB

JUVENILE

furry ruff on cheeks

grey-brown fur

long, bushy, tail

**SIZE** *Body 80–150cm; tail 30–50cm.*
**YOUNG** *Single litter of 3–6; March–May.*
**DIET** *Hunts deer, hares, rodents, and sometimes domestic stock; also carrion and berries, especially in the autumn.*
**STATUS** *Vulnerable (Italy), Conservation-dependent (Iberian Peninsula); locally common in E. and N.E. Europe.*
**SIMILAR SPECIES** *Golden Jackal (p.86) and domestic dogs.*

# Arctic Fox

*Alopex lagopus* (Carnivora)

**FOUND** *in Arctic tundra and on sea-ice, in mountains further south, and open woodland in winter.*

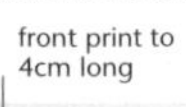

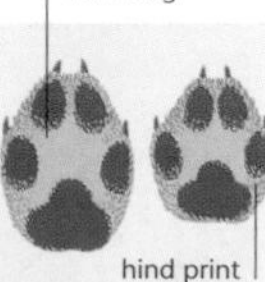


tapered droppings 4–8cm long

Small and dog-like, the Arctic Fox has grey or brown fur, usually turning white in the winter, which distinguishes it from the larger, redder Red Fox. One form known as the Blue Fox, which is predominant in Iceland but rare elsewhere, has a blue-grey winter coat. By way of adaptation to its cold northern habitats, its ears and muzzle are short and the coat is very thick, both serving to reduce heat loss.

SUMMER COAT

short ears

thick white (or blue-grey) winter coat

long bushy tail

**SIZE** *Body 50–75cm; tail 25–40cm.*
**YOUNG** *Single litter of up to 18.*
**DIET** *Lemmings, voles, birds and their eggs, shellfish, and carrion.*
**STATUS** *Locally common; rare where over-hunted for its fur.*
**SIMILAR SPECIES** *None.*

# Raccoon Dog

*Nyctereutes procyonoides* (Carnivora)

**OCCURS** *in extensive deciduous woodland, especially near lakes and rivers.*

A small, stocky, long-haired dog, the Raccoon Dog is solitary and nocturnal, and (uniquely in the dog family) hibernates. It has grey-brown fur, darker on the legs and on the face, where it forms a distinctive black mask. The mask is accentuated by a pale grey stripe running from above each eye to behind the ears. Introduced from Asia around 1930, the European population has expanded rapidly.

short ears, fringed with white

pale brow stripes

black face mask

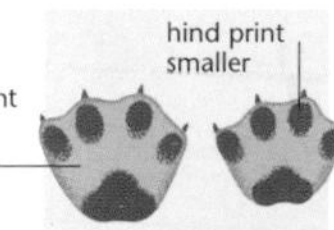


twisted droppings to 8cm long

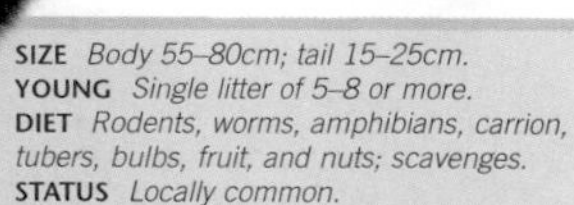

**SIZE** *Body 55–80cm; tail 15–25cm.*
**YOUNG** *Single litter of 5–8 or more.*
**DIET** *Rodents, worms, amphibians, carrion, tubers, bulbs, fruit, and nuts; scavenges.*
**STATUS** *Locally common.*
**SIMILAR SPECIES** *Raccoon (p.90); Badger (p.97).*

# Red Fox

*Vulpes vulpes* (Carnivora)

Worldwide the most abundant and widespread carnivore, the Red Fox, as it name suggests, is usually a red-brown colour, although this can vary from sandy yellow to dark brown. Its underparts are usually white or pale, as is the tip of its long, bushy tail. The lower part of the legs and the backs of the erect, triangular ears are blackish. Predominantly nocturnal and crepuscular, the Red Fox is increasingly seen by day in areas where it is not persecuted. It is most easily noticed by its droppings, the musky odour of its urine, or its nocturnal high-pitched barks and screams.

**OCCUPIES** *a vast range of habitats, from woodland and farms, to mountains and city centres; often secretive, but look out for its prominently displayed droppings.*

**NOTE**

*Versatile and adaptable, the Red Fox is the only large carnivore which has successfully colonized urban areas, finding food and excavating burrows (earths) in parks and gardens.*

**SIZE** *Body 55–90cm; tail 30–45cm.*
**YOUNG** *Single litter of 4–5; March–May.*
**DIET** *Omnivorous: rabbits, rodents, hedgehogs, birds and their eggs; beetles, worms, and other ground invertebrates; crabs; fruit and berries; carrion and food scavenged from refuse tips and bins.*
**STATUS** *Common.*
**SIMILAR SPECIES** *Golden Jackal (p.86).*

# Raccoon

*Procyon lotor* (Carnivora)

Introduced to Germany from North America in 1934, the Raccoon's combination of a black face mask and bushy tail, ringed with black and white, is unique. Largely nocturnal, Raccoons spend the day in a den in a tree-hole or rock crevice. They both swim and climb well, and exploit a wide range of foods accordingly, aided by their long front fingers which are adept at manipulating diverse food items.

**INHABITS** *woodland, especially close to water; has adapted to living in large parks and gardens, in close proximity to humans.*

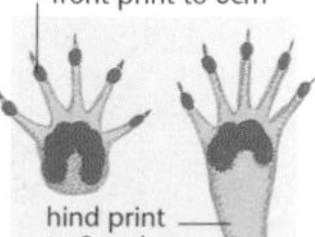


triangular ears with white margins

long, grey-brown fur

black face mask

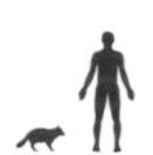

**SIZE** *Body 50–70cm; tail 20–25cm.*
**YOUNG** *Single litter of 2–5; March–May.*
**DIET** *Rodents, fish, shellfish, insects, fruit, and cereals; scavenges from refuse bins.*
**STATUS** *Scarce but increasing.*
**SIMILAR SPECIES** *Raccoon-dog (p.88); Badger (p.97).*

# Wolverine

*Gulo gulo* (Carnivora)

Also known as the Glutton, from its habit of gorging on available food during the winter months, the Wolverine is the largest member of the Weasel family. Heavily built, it resembles a small Brown Bear, but its pale neck and flank stripes are characteristic. Wolverines can climb trees, despite their bulk, but spend most of their time on the ground.

**ROAMS** *extensively around forests, bogs, tundra, and mountains; constructs dens in caves, rocks, or snow drifts.*

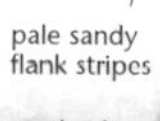

pale sandy flank stripes

long, bushy tail

dark brown fur

powerful legs

**SIZE** *Body 60–70cm; tail 15–25cm.*
**YOUNG** *Single litter of 2–3; February–March.*
**DIET** *Carrion, rodents, eggs, invertebrates, and fruit; mammals trapped in deep snow.*
**STATUS** *Vulnerable; rare, but currently increasing.*
**SIMILAR SPECIES** *A small Brown Bear (p.85).*

# Marbled Polecat

*Vormela peregusna* (Carnivora)

The Marbled Polecat is unique in its very dark brown underparts and its creamy yellow upperparts marbled with dark brown patches. Its long bushy tail is pale brown with a blackish tip. When threatened, this polecat lifts its tail over its back, displaying its contrasting colours as a warning. This species is mostly nocturnal.

OCCURS *in dry steppes, grassland, semi-desert habitats, and other dry areas, including cultivated land, scrub, and open, rocky woodland.*

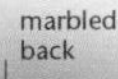

marbled back

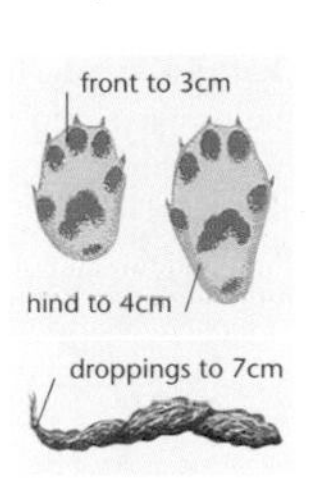

bicoloured face

droppings to 7cm

**SIZE** *Body 28–38cm; tail 12–22cm.*
**YOUNG** *Single litter of up to eight; January–March.*
**DIET** *Burrowing rodents, reptiles, ground-nesting birds, and insects.*
**STATUS** *Vulnerable; scarce and declining.*
**SIMILAR SPECIES** *None.*

# Steppe Polecat

*Mustela eversmanii* (Carnivora)

With the size, shape, and habits of the Western Polecat, the Steppe Polecat is distinguished by its more contrasting colour pattern – pale brown above but darker below – the sandy colour being typical of animals living in arid areas. Although it will make its own burrows, it often takes over those of sousliks or hamsters.

FREQUENTS *arid steppe grassland and semi-desert habitats, and cultivated fields, especially where there are colonies of its major prey species.*

pale sandy brown fur

white face

dark eye patches

front to 3cm

hind to 4cm

twisted droppings to 7cm

**SIZE** *Body 30–52cm; tail 10–18cm.*
**YOUNG** *Single litter of 3–6; April–May.*
**DIET** *Rodents, especially sousliks and hamsters, and large insects.*
**STATUS** *Locally common.*
**SIMILAR SPECIES** *Western Polecat (p.92), which is generally darker.*

# Western Polecat

*Mustela putorius* (Carnivora)

**OCCURS** *in lowland, often rocky woodland, river banks and marshes, sand dunes; also around farms and other buildings, particularly in the winter.*

The brown colour of this dark, slender predator is relieved only by its white face pattern, and where the outer coat is thinner on the flanks, the pale of the underfur may show through. As with most mustelids, it moves with a bounding gait, and readily adopts an alert posture, standing on its hind legs. Since it is largely nocturnal, it is not often seen, except all too often as a road casualty. However, its presence can be detected from the strong, musky scent it leaves as territorial marks on rocks and other landscape features.

**NOTE**

*The Ferret is the domesticated form of this Polecat, which often escapes from captivity. Its dark and pale forms are distinctive, but those with a polecat pattern can only be differentiated on skull characteristics.*

**SIZE** *Body 30–45cm; tail 12–14cm.*
**YOUNG** *Single litter of up to 12, May–July.*
**DIET** *Hunts rodents, rabbits, frogs, and birds; also takes insects, worms, and carrion.*
**STATUS** *Scarce, but widespread and increasing.*
**SIMILAR SPECIES** *Steppe Polecat* *(p.91)**, which is paler; minks* *(p.93)**, which are more uniformly dark; Ferret* (M. furo)*, which often escape.*

# European Mink

*Mustela lutreola* (Carnivora)

A nocturnal predator, the European Mink swims well although its feet are only partially webbed. It lacks the grace of an Otter, which floats low in the water with only its head exposed most of the time. In contrast, the European Mink floats high, with its head, back, and tail visible. It is a uniform dark brown, with a relatively long, bushy tail; the only colour relief is a small area of white fur on the chin, and upper and lower lips.

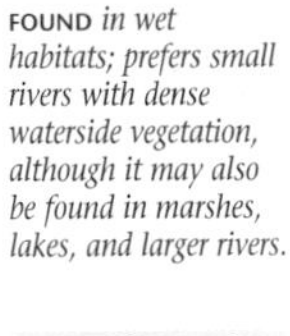

**FOUND** *in wet habitats; prefers small rivers with dense waterside vegetation, although it may also be found in marshes, lakes, and larger rivers.*

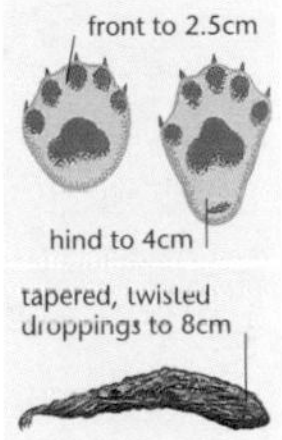


**SIZE** *Body 30–40cm; tail 12–15cm.*
**YOUNG** *Single litter of 4–5, April–June.*
**DIET** *Rodents, water birds, amphibians, fish, crayfish, and molluscs.*
**STATUS** *Endangered; rare and declining.*
**SIMILAR SPECIES** *American Mink (below); Western Polecat (p.92); Otter (p.96).*

# American Mink

*Mustela vison* (Carnivora)

Similar to its European relative, the American Mink is now the predominant species in Europe since its introduction in the 1920s. It is dark brown in colour but a number of colour variations have been bred on fur farms. The only way of separating the European and American species visually is by the lack of white on the upper lip of the American.

**INHABITS** *a variety of of waterside habitats, even rocky coasts; the droppings are similar to an Otter's but lack their "sweet" smell.*

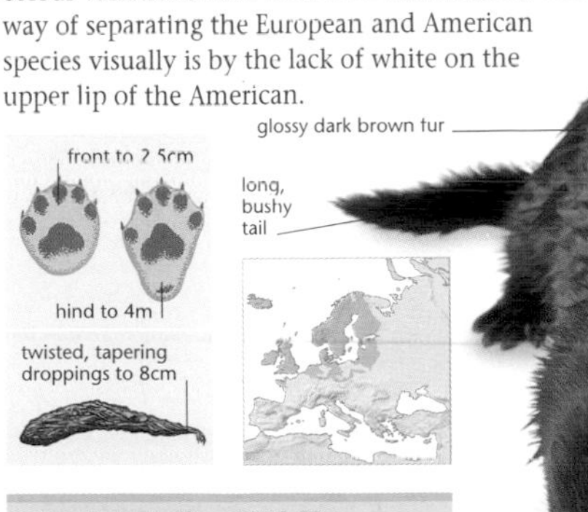

**SIZE** *Body 30–45cm; tail 13–20cm.*
**YOUNG** *Single litter of 4–6, April–May.*
**DIET** *Rodents, birds, amphibians, and fish.*
**STATUS** *Introduced from North America; locally common, and spreading.*
**SIMILAR SPECIES** *European Mink (above); Western Polecat (p.92); Otter (p.96)*

# Stoat

*Mustela erminea* (Carnivora)

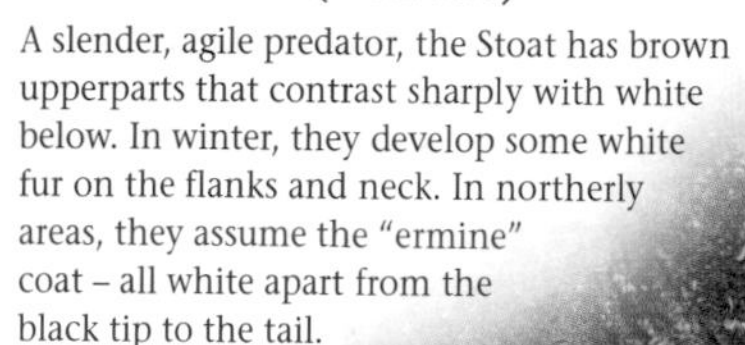

A slender, agile predator, the Stoat has brown upperparts that contrast sharply with white below. In winter, they develop some white fur on the flanks and neck. In northerly areas, they assume the "ermine" coat – all white apart from the black tip to the tail.

**FOUND** *in most terrestrial habitats with sufficient cover, from woodland and grassland, to marshes and mountains.*

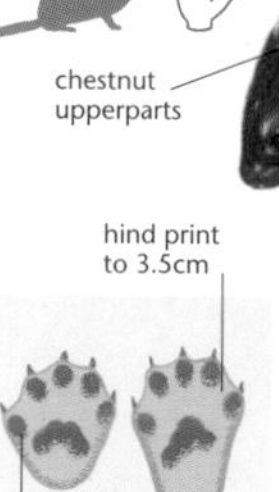


WINTER COAT

**SIZE** *Body 18–31cm; tail 9–14cm.*
**YOUNG** *Single litter of 6–12; April–May.*
**DIET** *Preys upon rodents (often pursued in their burrows), Rabbits, and birds.*
**STATUS** *Common.*
**SIMILAR SPECIES** *Weasel (below), which is smaller and does not have a black-tipped tail.*

# Weasel

*Mustela nivalis* (Carnivora)

Weasels are somewhat variable in size: males are smaller than females, and in the absence of Stoats, especially around the Mediterranean, they grow considerably larger than elsewhere. The tail is rather short, and does not have a black tip. Weasels from the far north (the so-called Least Weasel) and from some mountain areas turn white in winter; otherwise, the winter coat is the same colour as the summer one, although longer.

**OCCURS** *in all terrestrial habitats, from semi-desert to woodland, and coasts to mountain tops.*

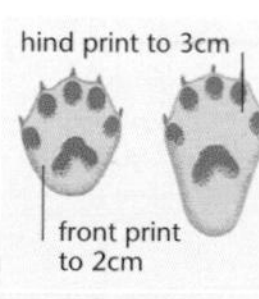

**SIZE** *Body 13–30cm; tail 3–10cm.*
**YOUNG** *Single or double litters of 4–6; April–August.*
**DIET** *Rodents, Rabbits, and birds; needs up to 10 meals each day.*
**STATUS** *Common.*
**SIMILAR SPECIES** *Stoat (above).*

# Pine Marten

*Martes martes* (Carnivora)

The Pine Marten has a sinuous body shape, long legs, and a long bushy tail. It is well-adapted to hunting in trees, in which it climbs with agility despite its relative bulk. Largely nocturnal, it makes its den in holes and crevices in rocks and trees. Its most distinctive feature is the pale throat patch, which varies in colour, from cream to pale orange.

**LIVES** *mainly in woodland (especially coniferous and mixed), but also colonizes scrub, rocky, and cliff habitats.*

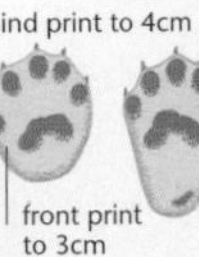


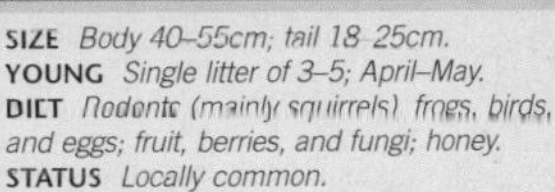

**SIZE** *Body 40–55cm; tail 18–25cm.*
**YOUNG** *Single litter of 3–5; April–May.*
**DIET** *Rodents (mainly squirrels), frogs, birds, and eggs; fruit, berries, and fungi; honey.*
**STATUS** *Locally common.*
**SIMILAR SPECIES** *Stone Marten (below), which is heavier built.*

# Stone Marten

*Martes foina* (Carnivora)

Also known as the Beech Marten, the Stone Marten usually has a distinct white throat patch, but this is very variable in extent. Indeed, in some areas such as Crete, the throat patch is virtually absent. A nocturnal predator, it makes dens in rock crevices, and frequently in buildings, where it often damages electrical wiring by chewing it.

**INHABITS** *a wide range of habitats, including rocky and mountainous areas, deciduous woodland, farmland, and towns.*

hind print to 4cm
front print to 3cm
droppings to 10cm long

**SIZE** *Body 40–48cm; tail 22–25cm.*
**YOUNG** *Single litter of 3–4; March–May.*
**DIET** *Rodents, shrews, and small birds; fruit and berries in the autumn.*
**STATUS** *Common and increasing.*
**SIMILAR SPECIES** *Pine Marten (above); Sable* (M. zibellina).

# Otter

*Lutra lutra* (Carnivora)

**FAVOURS** *rivers, lakes, estuaries, and sheltered rocky coasts, without significant disturbance or pollution.*

A predominantly aquatic mammal, the Otter is generally seen in or near water. However, it will travel long distances overland at night, its main activity period: at such times, otters are frequently killed on roads. On land, it travels with an awkward, bounding gait, but in water it is very agile and playful. It swims low in the water, with just the head exposed, lower than any of the minks or large aquatic rodents with which it may be confused. When underwater, it has a silvery appearance due to bubbles of air trapped in the fur. It has whitish underparts and small ears.

sleek, brown fur

**NOTE**

*Given their secretive nature, the best way to establish the presence of otters is to search for their spraints – oily, sweet-smelling droppings, usually containing fish scales and bones, mostly near water.*

**SIZE** *Body 55–90cm; tail 35–50cm.*
**YOUNG** *Single litter of 2–5; May–August, but can be all year round.*
**DIET** *Fish, amphibians, rodents, water birds, and crustaceans; crabs and molluscs in coastal areas; carrion.*
**STATUS** *Scarce, though locally common.*
**SIMILAR SPECIES** *Large aquatic rodents – beavers (pp.61–62); Coypu (p.62), and Muskrat (p.63); European and American Minks (p.93).*

# Badger

*Meles meles* (Carnivora)

The Badger is largely nocturnal and, therefore, rarely encountered, except as road casualties, unless actively sought out. However, its complex burrow system (set) with an often extensive series of holes 20cm or more across is very obvious, and forms the focus of most surveys for the species. The characteristic black and white facial stripes are surprisingly visible even in twilight conditions, when most badgers are observed. Heavily built, the Badger has a small pointed head and short neck, widening to a powerful body, with short, strong limbs and a small tail.

**OCCURS** *in a range of habitats from lowland to subalpine levels; setts are usually in the cover of woodland, scrub, or hedges, but forages in open areas.*

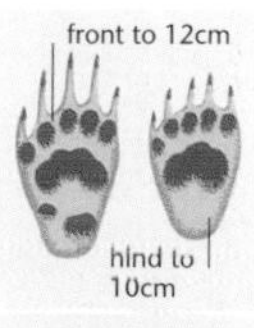


brown droppings to 10cm

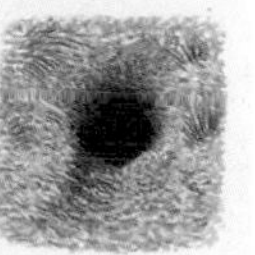


grey upperparts

small, white-fringed ears

black and white striped head

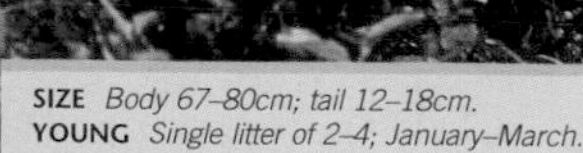

**SIZE** *Body 67–80cm; tail 12–18cm.*
**YOUNG** *Single litter of 2–4; January–March.*
**DIET** *Omnivorous; earthworms, insects, small mammals, amphibians, bulbs, grain, and carrion.*
**STATUS** *Common.*
**SIMILAR SPECIES** *Raccoon-dog (p.88) and Raccoon (p.90), which have black and white faces but differ in size, shape, and habit.*

# Genet

*Genetta genetta* (Carnivora)

**FAVOURS** *open woodland and rocky scrub, up to high levels in the mountains and often around rivers and streams*

A rather short-legged, cat-like animal, related to the mongooses, the most distinctive feature of the Genet is its very long tail, ringed with black and white. Its upperparts are a pale sandy colour, spotted with black, and it has large ears. An agile climber, the Genet is largely nocturnal. It often roosts in tree tops during the day, and breeds in holes in trees and rocks.

large ears

spots arranged in broken stripes

long, ringed tail

footprints to 3cm long

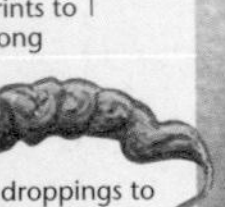

droppings to 2.5cm long

**SIZE** *Body 50–60cm; tail 40–50cm.*
**YOUNG** *One or two litters of 2–4.*
**DIET** *Rodents, rabbits, birds; reptiles, insects, and berries.*
**STATUS** *Vulnerable; generally scarce.*
**SIMILAR SPECIES** *Wild Cat (p.99) and domestic cats.*

# Egyptian Mongoose

*Herpestes ichneumon* (Carnivora)

**PREFERS** *scrub and rocky slopes; also found by rivers and watercourses.*

Distinctive in its grizzled grey-brown coat, the Egyptian Mongoose resembles polecats and martens in its slim build, but has rather longer legs and a tapering tail. Mainly nocturnal, it often lives in rabbit warrens. Renowned for killing and eating snakes, it displays some resistance to the bites of venomous species.

**SIZE** *Body 50–55cm; tail 35–35cm.*
**YOUNG** *Single litter of 2–4.*
**DIET** *Rabbits, rodents, birds, reptiles, insects.*
**STATUS** *Probably introduced; locally common.*
**SIMILAR SPECIES** *Polecats (pp.91–92); Minks (p.93).*

footprints to 3cm long

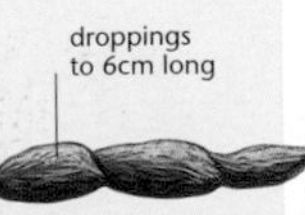

droppings to 6cm long

# Wild Cat

*Felix sylvestris* (Carnivora)

The size of a large domestic cat, Wild Cats can be very difficult to tell apart from feral tabbies and especially hybrids. Solitary and nocturnal, the Wild Cat is an agile climber, although it hunts mainly on the ground. Despite stalking and pouncing in the familiar cat manner, it shows no tendency to "play" with its prey. It is short-legged, and so is unable to tolerate snow deeper than 20cm for long periods.

**INHABITS** *woodland, scrub, and rocky areas, mainly in mountains; makes dens in hollow trees, rock crevices, and the burrows of other mammals.*

**NOTE**

*Definitive separation from some domestic cats, and especially hybrids is very difficult, and may rely on skull measurments.*

yellow-green eyes

dark markings forming stripes, not blotches

black-tipped, bushy tail with 3–5 dark rings

DOMESTIC TABBY

**SIZE** *Body 48–65cm; tail 20–35cm.*
**YOUNG** *Single litter of 2–4, April-September.*
**DIET** *Rabbits, hares, rodents, birds, lizards and frogs; exceptionally takes lambs.*
**STATUS** *Vulnerable (Scottish form); scarce and local, though some recent signs of increase; threatened by interbreeding with domestic cats.*
**SIMILAR SPECIES** *Genet (p.98); domestic cats and hybrids.*

footprints to 4cm long

droppings to 5cm long

**ROAMS** *widely in conifer forests and mountains; its long legs enable it to move in deep snow.*

# Lynx

*Lynx lynx* (Carnivora)

The largest cat in Europe, the Lynx is recognizable by its tufted ears and cheeks, short tail, long legs, and usually spotted coat, particularly on its legs; animals from northern areas are only weakly spotted. Rarely seen, it is solitary and nocturnal. Lynx have been hunted to extinction over much of western Europe.

prints to 10cm

no claw marks

droppings to 8cm long

**SIZE** *Body 0.8–1.3m; tail 10–25cm.*
**YOUNG** *Single litter of 2–3; May–June.*
**DIET** *Hares, small deer, rodents, and game birds.*
**STATUS** *Rare or very locally scarce; some reintroduction schemes are taking place.*
**SIMILAR SPECIES** *Iberian Lynx (below).*

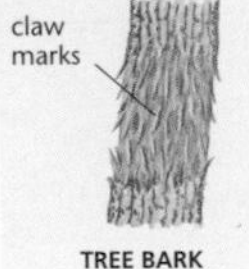


**TREE BARK**

**FOUND** *in open Mediterranean woodland and scrub, rocky habitats, and sand dunes; avoids cultivated areas.*

# Iberian Lynx

*Lynx pardinus* (Carnivora)

A little smaller and more densely spotted than the Lynx, the Iberian Lynx, also known as the Pardel Lynx, is one of the rarest European mammals. Geographically separated from the Lynx, it is found in about 12 isolated populations, all small, and so exposed to the risks of inbreeding.

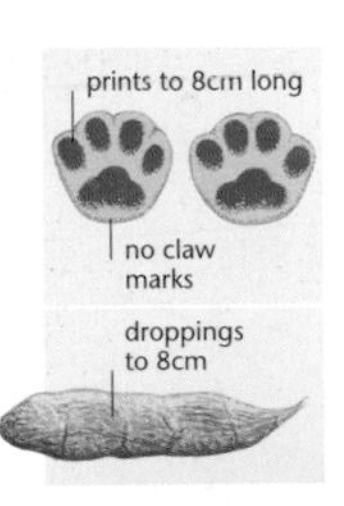


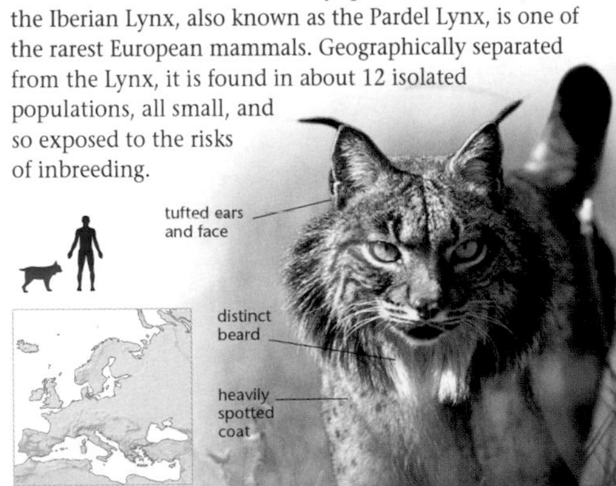


**SIZE** *Body 80–110cm; tail 10–15cm.*
**YOUNG** *Single litter of 2–3; May–June.*
**DIET** *Rabbits, hares, young deer, and ground-dwelling birds.*
**STATUS** *Endangered; rare and declining as a result of habitat loss and fragmentation.*
**SIMILAR SPECIES** *Lynx (above).*

# Common Seal

*Phoca vitulina* (Pinnipedia)

Also known as the Harbour Seal, the Common Seal is one of the smaller members of its family, with a grey or brownish coat, variably speckled with black. Its short muzzle protrudes from the forehead, and with its white whiskers, gives it a dog-like profile. Its nostrils form a V-shaped pattern. Pups are born with a similar colour to their parents, and they are capable of swimming almost immediately after birth. Common Seals can, therefore, breed on tidal flats.

**OCCUPIES** *shallow coastal waters and estuaries, sometimes swimming great distances up rivers; seen on rocks and sand banks at low tide.*

**NOTE**

*Over most of the northwest coast of Europe, only Common and Grey Seals are likely to be found; they can be distinguished by size and their facial profiles, which are "dog-like" and "Roman-nosed", respectively.*

JUVENILE

**SIZE** *Body 1.2–1.9m; females a little smaller than males.*
**YOUNG** *Single pup; June–July.*
**DIET** *Fish, shellfish, molluscs, and crustaceans.*
**STATUS** *Locally common; numbers still recovering from recent viral epidemics.*
**SIMILAR SPECIES** *Other seals, especially Ringed Seal (p.102), although the coat of the latter is usually marked with pale rings.*

# Ringed Seal

*Phoca hispida* (Pinnipedia)

**SWIMS** *in inshore water, bays, and fjords, and breeds on ice sheets; readily enters fresh water.*

The smallest European seal, with males and females similar in size, the Ringed Seal usually has a dark grey-brown back. Like the Common Seal, it has a dog-like, concave facial profile. It is the commonest Arctic seal, breeding also in the Baltic, and rarely wanders far from its breeding areas; as a result it is rarely or never seen around most of the European coastline. Pups are white and woolly at birth.

**SIZE** *Body 1.2–1.35m.*
**YOUNG** *Single litter of 1–2; February–April.*
**DIET** *Fish and small crustaceans.*
**STATUS** *Endangered and Vulnerable (Baltic and lake populations respectively); otherwise locally very common.*
**SIMILAR SPECIES** *Common Seal (p.101).*

# Harp Seal

*Phoca groenlandica* (Pinnipedia)

**INHABITS** *Arctic pack ice (on which it breeds) and open ocean; moves with the seasonal expansion and contraction of the ice edge.*

With its striking black and white pattern in the shape of a harp, a male Harp Seal is unmistakable; even females have a shadowy trace of the male pattern and are readily recognizable. Harp Seals are Arctic residents, breeding in dense colonies in spring on pack ice. Pups, born white, are a target for hunters. After about two weeks of age, they moult to a greyish colour and are deserted by their parents.

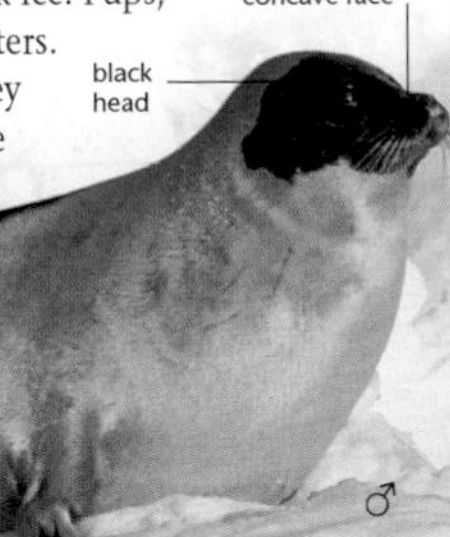

**SIZE** *Body 1.7–1.9m.*
**YOUNG** *Single pup; February–March.*
**DIET** *Fish, especially capelin and cod, and crustaceans.*
**STATUS** *Locally very common; they have long been hunted for their fur.*
**SIMILAR SPECIES** *None.*

# Walrus

*Odobenus rosmarus* (Pinnipedia)

The largest of the European pinnipeds, the prominent tusks of the Walrus make it easy to identify. Those of the male are especially impressive, up to 70cm long. The Walrus used its tusks to haul itself onto ice, to dredge molluscs as food, in social displays, to kill seals and to defend against Polar Bears. It breeds in huge herds; the pups are born dark grey, before moulting into a brown coat.

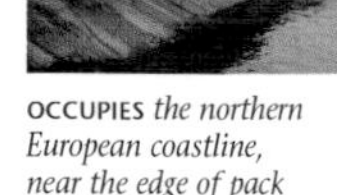

**OCCUPIES** *the northern European coastline, near the edge of pack ice on which it breeds.*

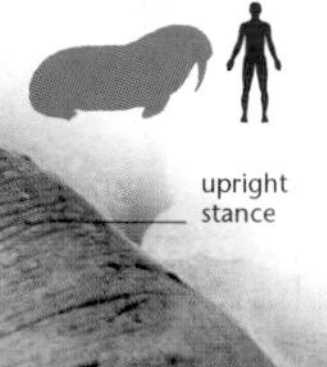


**SIZE** *Males 2.8–3.6m; females 2.5–3m.*
**YOUNG** *Single litter of 1–2; April–June: females breed only every two or three years.*
**DIET** *Molluscs, seals, crustaceans, and fish.*
**STATUS** *Locally common; still recovering from near-extinction in the 1950s.*
**SIMILAR SPECIES** *None.*

# Mediterranean Monk Seal

*Monachus monachus* (Pinnipedia)

The only seal in the Mediterranean and Black Seas, the Mediterranean Monk Seal is also found in northwest Africa and Madeira. However, it is rare in all these areas; the world population is thought to number only some 300–400 individuals. It is a large seal, the sexes of a similar size, with rich chestnut brown upperparts and a variable amount of white beneath. The pups are black at birth.

**LIVES** *along coastlines; breeds in sea caves or secluded beaches; feeds in adjacent relatively shallow water; highly intolerant of disturbance.*

**SIZE** *Body 2.3–2.8m.*
**YOUNG** *Single pup; September–October, though not every year.*
**DIET** *Fish and marine invertebrates.*
**STATUS** *Critically Endangered; rare and extinct over large parts of former range.*
**SIMILAR SPECIES** *None.*

# Grey Seal

*Halichoerus grypus* (Pinnipedia)

♂

**FOUND** *along coasts and in coastal marine waters, breeding on rocky islets, grassy coastal strips, and ice shelves in the north.*

Male Grey Seals are very large, about a third larger than the females, and both sexes have variably blotchy grey upperparts and paler undersides. In profile, the forehead runs straight into the muzzle and the nostrils are widely separated. Pups are born white and remain on land for several weeks – Grey Seals, therefore, must breed above the high tide mark. In water, they are capable of diving to depths of greater than 200m, remaining submerged for up to 30 minutes at a time.

**NOTE**

*The long, straight facial profile of this seal is especially characteristic of males; females and juveniles have a shorter muzzle, but never show the concave profile of most other European seal species.*

grey above, variably blotched

silvery below, often blotched

♀

# Minke Whale

*Balaenoptera acutorostrata* (Cetacea)

**INHABITS** *the open sea, especially in shallower continental shelf waters.*

The smallest of the baleen whales, the Minke Whale is also the most visible from land. Often all that can be seen is its curved dorsal fin, which is relatively larger than the dorsal fins of other baleen whales. When a Minke Whale dives, its pointed snout breaks the surface, but the tail flukes remain hidden. Occasionally, it breaches, revealing characteristic white patches on the flippers.

up to 2m high

BLOW

pointed snout

flattened head

**SIZE** *8–10.5m.*
**YOUNG** *Single young; December–January.*
**DIET** *Fish, crustaceans, and squid.*
**STATUS** *Near-threatened; generally scarce, though locally common.*
**SIMILAR SPECIES** *Fin Whale* (B. physalis), *is larger with asymmetric head markings.*

**SIZE** *Males 2.2–3m; females 2–2.5m.*
**YOUNG** *Single pup; normally October–February.*
**DIET** *Fish (including commercial species such as Salmon), crustaceans, and cephalopods.*
**STATUS** *Endangered (Baltic population only); locally common.*
**SIMILAR SPECIES** *Hooded Seal* (Cystophora cristata), *which has larger, irregular dark blotches; males have inflatable hoods.*

JUVENILE

# Humpback Whale

*Megaptera novaeangliae* (Cetacea)

The Humpback Whale is one of the easier whales to identify, especially when it displays its striking dive sequence – the steepness of the dive is such that the tail flukes are raised out of the water, revealing white undersides. Furthermore, Humpbacks also breach regularly, when the very long, whitish flippers may be seen.

*FEEDS in shallow seas; breeds in the Tropics, moving to Arctic waters in the summer.*

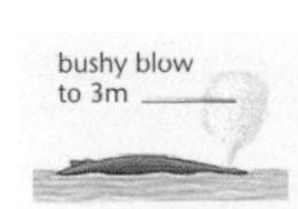


BLOW

**SIZE** *11–18m.*
**YOUNG** *Single young; any female breeds only every two or three years.*
**DIET** *Small schooling fish and large invertebrates.*
**STATUS** *Vulnerable; scarce to rare inshore.*
**SIMILAR SPECIES** *None.*

# Harbour Porpoise

*Phocoena phocoena* (Cetacea)

**OCCURS** *in shallow coastal waters and estuaries right around Europe, although scarce in the Mediterranean; substantial aggregations can form in favoured feeding areas.*

The commonest (and smallest) cetacean in European waters, the Harbour Porpoise is relatively nondescript, steely grey above and whitish below, the pigmentation usually being asymmetric. The dorsal fin, often the only feature seen on a live animal, is short and blunt, and located centrally down the back. It has a rounded head and spade-shaped teeth. Unlike most dolphins, the snout is not extended into a beak; nor is it typically as agile as a dolphin, as Harbour Porpoises do not leap clear of the water.

**NOTE**

*Weather conditions influence the likelihood of viewing small cetaceans such as Harbour Porpoises; the merest hint of waves easily disguise the occasional fleeting dorsal fin.*

# Atlantic White-sided Dolphin

*Lagenorhynchus acutus* (Cetacea)

**OCCURS** *in temperate and sub-Arctic waters, usually along the edge of the continental shelf, but is sometimes found inshore.*

A large, short-beaked dolphin, the Atlantic White-sided Dolphin has three distinctive features: its black beak, pale flank line, and its tail, which is deep but narrow and vertically flattened just in front of the flukes. The black, sickle-shaped dorsal fin is noticeably long in comparison wtih other species.

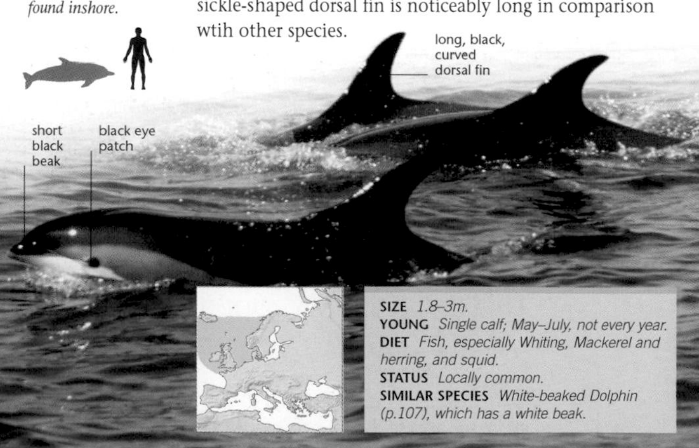


**SIZE** *1.8–3m.*
**YOUNG** *Single calf; May–July, not every year.*
**DIET** *Fish, especially Whiting, Mackerel and herring, and squid.*
**STATUS** *Locally common.*
**SIMILAR SPECIES** *White-beaked Dolphin* *(p.107)**, which has a white beak.*

**SIZE** *1.4–1.8m.*
**YOUNG** *Single young; born May–August; not every year.*
**DIET** *Small fish, especially Herrings; also some crustaceans and cuttlefish.*
**STATUS** *Vulnerable; locally common, although Baltic and Mediterranean populations are threatened.*
**SIMILAR SPECIES** *None.*

# White-beaked Dolphin

*Lagenorhynchus albirostris* (Cetacea)

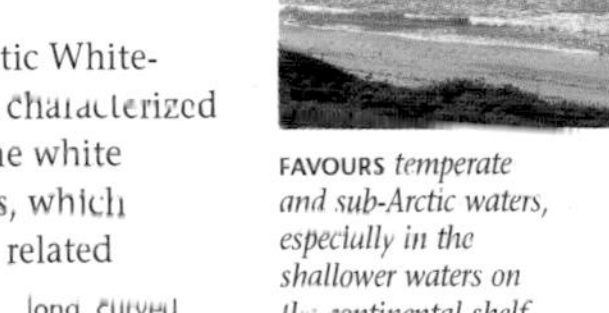

**FAVOURS** *temperate and sub-Arctic waters, especially in the shallower waters on the continental shelf.*

Large and short beaked, rather like the Atlantic White-sided Dolphin, the White-beaked Dolphin is characterized by its white beak, the white connecting to the white underparts. It has pale patches down its sides, which extend further forward on the body than on related species. Its dorsal fin is long, erect, and curved. It is less sociable than the Atlantic White-sided Dolphin, but very agile and sometimes bow-rides with boats.

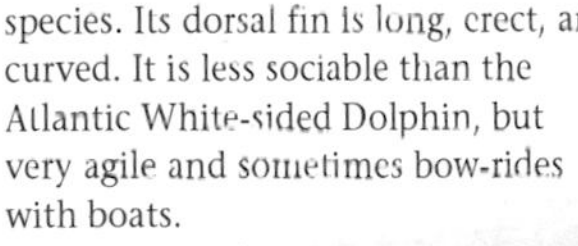

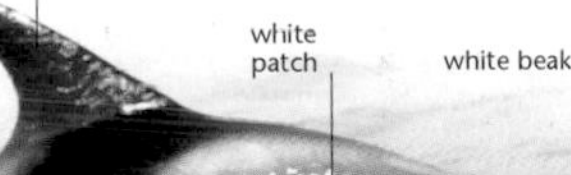

**SIZE** *2.3–2.7m.*
**YOUNG** *Single calf; May–August.*
**DIET** *Fish, especially Cod and Whiting, and octopus.*
**STATUS** *Locally common.*
**SIMILAR SPECIES** *Atlantic White-sided Dolphin (p.106).*

# Long-finned Pilot Whale

*Globicephala melas* (Cetacea)

**FAVOURS** *shallow seas and deeper oceanic waters, from the Arctic to the Mediterranean.*

With a bulbous forehead, the Long-finned Pilot Whale is a distinctive, very large member of the dolphin family. Its blackish upperparts, recurved dorsal fin, and long, slender flippers are often easy to see when groups are resting on the surface in calm sea conditions. Pilot Whales are very gregarious, often in groups of a hundred or more, and prone to mass strandings.

sickle-shaped dorsal fin

bulbous forehead

**SIZE** *5–6.5m.*
**YOUNG** *Single calf; July–September.*
**DIET** *Cuttlefish, squid, fish, and crustaceans.*
**STATUS** *Locally common.*
**SIMILAR SPECIES** *Risso's Dolphin (p.109); Bottle-nosed Whale* (Hyperodon ampullatus), *which is paler.*

# Killer Whale

*Orcinus orca* (Cetacea)

**INHABITS** *deeper waters, although regularly seen inshore where the water is deep enough, as in the northern fjords.*

The largest dolphin, with unmistakable black and white pattern, the Killer Whale has an extremely long, erect, triangular dorsal fin, to 1.8m in an adult male. The fins of females and juveniles are shorter and often recurved. Killer Whales are usually found in small groups, although more substantial numbers may be found around colonies of seals, one of their favoured foods.

large dorsal fin

large, paddle-shaped flippers

black and white colour

**SIZE** *5–9.5m.*
**YOUNG** *Single calf; born October–January.*
**DIET** *Seals, small dolphins and Harbour Porpoises; large fish, and carrion.*
**STATUS** *Scarce.*
**SIMILAR SPECIES** *Some sharks with a vertical tail fin.*

# Risso's Dolphin

*Grampus griseus* (Cetacea)

A bulky species with a rather bulbous forehead, Risso's Dolphin is usually a pale greyish colour above, although this can fade with age; the skin is frequently marked with extensive pale scars. Its dorsal fin is relatively long and slender, with a marked recurve. Very sociable, it forms large herds of a hundred or more, and regularly follows ships, often in the company of other cetaceans.

**PREFERS** *warm seas; usually to be found in deep water, though regularly seen from suitable vantage points on land.*

pale grey, scarred skin

slender, recurved dorsal fin

bulbous forehead

large eyes, separated by a deep furrow

**SIZE** *3.3–3.8m.*
**YOUNG** *Single calf; born April–September.*
**DIET** *Squid, octopus, cuttlefish, and small fish.*
**STATUS** *Scarce, though very locally common.*
**SIMILAR SPECIES** *Long-finned Pilot Whale (p.108), which is larger and blacker.*

# Bottlenose Dolphin

*Tursiops truncatus* (Cetacea)

The most widespread and numerous dolphin over much of the European coastline, the Bottle-nosed Dolphin has greyish upperparts, a short beak, and a long, curved dorsal fin. Despite its bulky appearance, it is very acrobatic, often leaping clear of the water, and it can be very inquisitive, approaching swimmers and boats closely.

**FOUND** *worldwide in tropical and temperate seas, from shallow estuaries to deep oceanic water.*

grey-brown upperparts

short but distinct beak

large, recurved dorsal fin

**SIZE** *2.5–4m.*
**YOUNG** *Single calf; born usually April–September, at intervals of up to three years.*
**DIET** *Fish, cuttlefish, and squid.*
**STATUS** *Locally common.*
**SIMILAR SPECIES** *None.*

# Wild Boar

*Sus scrofa* (Artiodactyla)

**INHABITS** *deciduous and coniferous forest with dense scrub and undergrowth; moves into marshes and cultivated areas to feed.*

The wild ancestor of the domestic pig, the Wild Boar shares many features characteristic of the farm animal – a heavy body, short neck and tail, large head, and elongated snout. In addition, it is covered in a coat of bristly hairs, which becomes darker and thicker in winter. Its whole body is flattened vertically, an adaptation to running between trees in its woodland habitats. In contrast, the piglets are horizontally striped with pale lines. Largely nocturnal, males are solitary, whereas the females form family groups with their litter.

**NOTE**

*Wild Boar are rather secretive, but leave copious field signs: cloven-hoofprints, rooting areas, wallows, scratching posts, and large cylindrical droppings.*

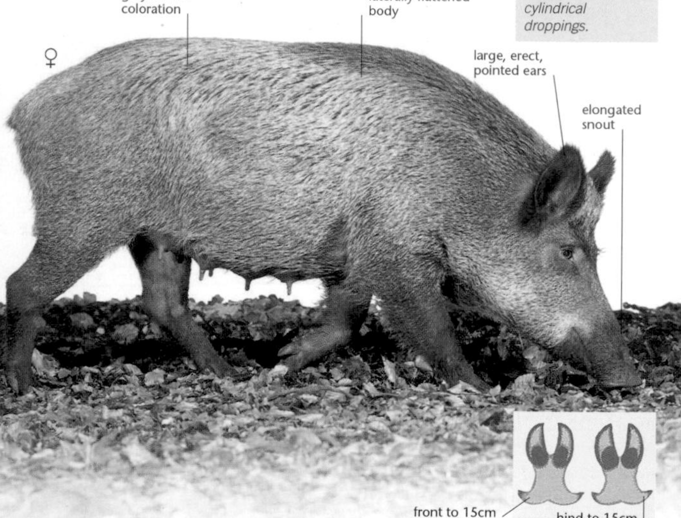


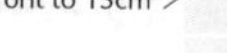

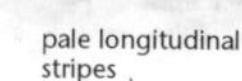

**PIGLETS**

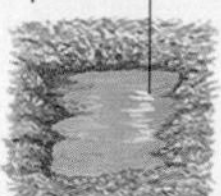


**MUD WALLOW**

**SIZE** *Body 1.1–1.7m; tail 13–30cm.*
**YOUNG** *Single litter of 2–10; February–June.*
**DIET** *Omnivorous: bulbs, tubers, acorns, beechmast, fruit, crops, fungi, insects, small mammals, ground-nesting birds, and carrion.*
**STATUS** *Locally common; sometimes as a pest population.*
**SIMILAR SPECIES** *Domestic pigs; hybridization produces animals which are intermediate in appearance, but reputedly more aggressive.*

# European Bison

*Bison bonasus* (Artiodactyla)

Reintroduced from captive stock following its extinction in the wild in the early 20th century, the European Bison resembles a hump-backed domestic cow, with its dark brown fur developed into a mane, especially in the bulls. Mainly nocturnal, the bulls are solitary and the cows move in herds, coming together only for the rut in August.

**FAVOURS** *extensive deciduous forests, with open glades for grazing; also in mountain pastures and marshland.*

front to 10cm
hind to 10cm
fibrous droppings

**SIZE** *Body 2.5–2.7m; tail 60–90cm.*
**YOUNG** *Single calf; May–June.*
**DIET** *Trees, shrubs, grasses and herbs; also acorns, beechmast, and bark in winter.*
**STATUS** *Endangered; rare.*
**SIMILAR SPECIES** *Some domestic breeds of cattle are quite similar.*

# Musk-ox

*Ovibos moschatus* (Artiodactyla)

A small cow, the Musk-ox is covered in long, dark fur, which in winter can reach down almost to the hooves, a necessary defence against Arctic weather. Both sexes bear curved horns, the bases of which touch on the crown. Present in Europe before the end of the last Ice Age, it then became extinct until introduced to Norway in the mid-20th century.

**OCCURS** *in montane tundra, moving down into open birch forests to protect itself from the elements in winter*

**SIZE** *Body 1.9–2.5m; tail 10–15cm.*
**YOUNG** *Single calf (occasionally twins); April–June; breeds usually every other year.*
**DIET** *Grazes and browses on tundra plants and shrubs.*
**STATUS** *Introduced; rare.*
**SIMILAR SPECIES** *None.*

# Mouflon

*Ovis ammon* (Artiodactyla)

**FAVOURS** *scrub, rocky areas, and open woodland, from lowland up to and around the tree-line in the mountains.*

The male of this small, short-haired species of sheep bears a pair of magnificent curved horns and has a distinctive dark chestnut brown coat, white rump, and white lower legs. Females, in contrast, are generally without horns, are a paler brown, with strikingly white underparts and with a black and white rump and tail. Mouflon are generally nocturnal, especially where hunted. Their complex history and links with domestic sheep have led to a confusing array of different colours, sizes, and horn shapes.

white eye rings

large coiled horns

white saddle-patch

white below

♀

♂

short, dark chestnut brown fur

white socks

prints to 5cm long

oval droppings to 1cm long

**SIZE** *Body 1.1–1.2m; tail 6–10cm.*
**YOUNG** *Single litter of 1–2; April–May.*
**DIET** *Grasses, sedges, and herbs; browses bark, shoots, and lichens.*
**STATUS** *Vulnerable (Mediterranean island populations); introduced widely elsewhere.*
**SIMILAR SPECIES** *Females resemble Chamois (pp.114–115), but lack the distinctive horns and face pattern; some breeds of domestic sheep.*

**NOTE**

*Now widely distributed in Europe, Asia, and the Americas, Mouflon probably originated in southwest Asia. Populations in Corsica, Sardinia, and Cyprus are believed to be ancient introductions.*

# Alpine Ibex

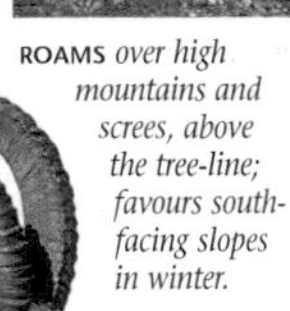

*Capra ibex* (Artiodactyla)

A montane wild goat, the male Alpine Ibex has unmistakable massive horns. The female, in contrast, has short horns and may be confused with Alpine Chamois, which occur in similar habitats. A very agile climber, this species moves with ease in rocky areas.

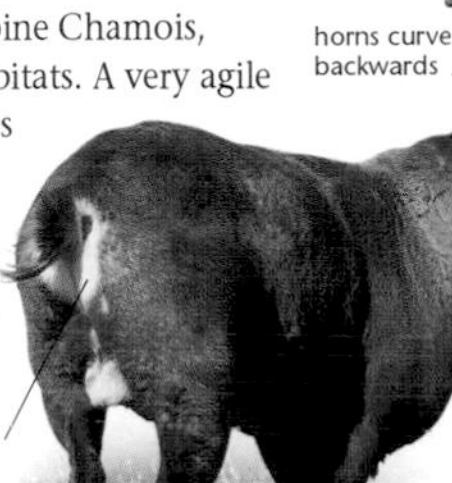


**ROAMS** *over high mountains and screes, above the tree-line; favours south-facing slopes in winter.*

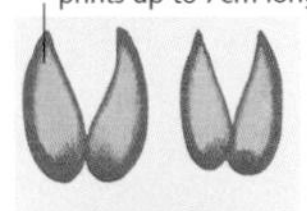

**SIZE** *Body 1.3–1.5m; tail 12–15cm.*
**YOUNG** *Single litter of 1–2; April–May.*
**DIET** *Grasses, herbs, shrubs, and lichens.*
**STATUS** *Scarce; very locally common.*
**SIMILAR SPECIES** *Spanish Ibex (below); Wild Goat (p.114); Alpine Chamois (p.115), which has a black and white face pattern.*

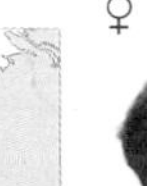

# Spanish Ibex

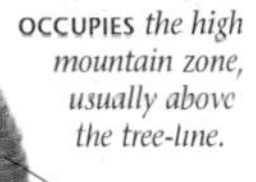

*Capra pyrenaica* (Artiodactyla)

The wild mountain goat of most Spanish mountain ranges, including the Pyrenees, the Spanish Ibex has horns that are are curved back, flared outwards, and are not prominently ribbed. The coat has colour contrasts, but this and the horns are variable between populations.

**OCCUPIES** *the high mountain zone, usually above the tree-line.*

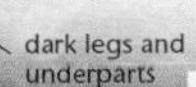

**SIZE** *Body 1.2–1.5m; tail 12–15cm.*
**YOUNG** *Single litter of 1–2; April–May.*
**DIET** *Montane grasses, herbs, dwarf shrubs, and lichens.*
**STATUS** *Critically Endangered; rare.*
**SIMILAR SPECIES** *Alpine Ibex (above), Wild Goat (p.114), and Southern Chamois (p.114).*

# Wild Goat

*Capra aegagrus* (Artiodactyla)

**FAVOURS** *arid, Mediterranean rocky habitats, scrubby areas, and mountains, where it lives in loose flocks.*

The ancestor of the domestic goat, the Wild Goat is predominantly an Asian species, but small populations are found on some Greek islands. A pure-bred male Wild Goat should have horns sweeping back in a gentle curve, diverging a little at the tips, a keel down the front, and a few prominent, but widely spaced ribs.

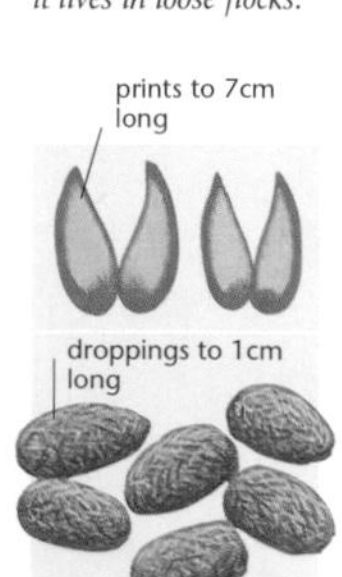

**SIZE** *Body 1–1.5m; tail 10–15cm.*
**YOUNG** *Single litter of 1–2; April–May.*
**DIET** *Browses trees and shrubs.*
**STATUS** *Vulnerable; generally scarce.*
**SIMILAR SPECIES** *Alpine Ibex and Spanish Ibex (p.113), both of which are geographically distinct.*

# Southern Chamois

*Rupicapra pyrenaica* (Artiodactyla)

**FOUND** *in high mountain pastures, rocks, and screes; moving into lower areas in heavy snow.*

Often known by its local name of Isard, the Southern Chamois forms three discrete populations in the Pyrenees, northwest Spain, and the Apennines. It resembles the Alpine Chamois in size and shape, with its hooked horns (in both sexes), and a black and white face. A patch of pale fur on the side of the neck gives a colour contrast not present in the Alpine species.

pale brown summer coat

pale throat and neck patches

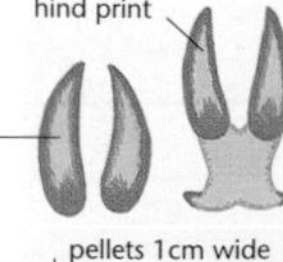

**SIZE** *Body 1–1.2m; tail 7–8cm.*
**YOUNG** *Single litter of 1–2; May–June.*
**DIET** *Alpine plants, grasses, and shrubs.*
**STATUS** *Endangered (Apennine population); very locally common in the other areas.*
**SIMILAR SPECIES** *Spanish Ibex (p.113); Alpine Chamois (p.115).*

# Alpine Chamois

*Rupicapra rupicapra* (Artiodactyla)

A small, goat-like animal found in mountain areas, the Alpine Chamois has a uniform coat, pale brown in the summer, but darker and woolly in the winter. Its black-and-white facial pattern is visible at a surprisingly long range. The hooked horns of both sexes are very distinctive. A gregarious species, the Alpine Chamois forms small groups of 20 or so in the summer, which in winter merge into larger herds.

**INHABITS** *mountain pastures, screes, and rocks, though moving lower in winter; its agility on cliffs and rocks is due partly to its flexible hooves with suction-like pads.*

**NOTE**

*In a vast open mountain-scape, a few Chamois can easily be overlooked; a good tip is to scan over distant snow-beds, remote from disturbance, where they may often be seen silhouetted against the snow.*

**SIZE** *Body 1–1.3m; plus tail 7–8cm.*
**YOUNG** *Single litter of 1–2; May–June.*
**DIET** *Grazes and browses on grasses, herbs, shrubs, and trees.*
**STATUS** *Two outlying populations are considered Critically Endangered; otherwise locally common.*
**SIMILAR SPECIES** *Ibexes (p.113), which are larger and heavier built, with larger horns; Southern Chamois (p.114), which has more contrast.*

# Elk

*Alces alces* (Artiodactyla)

**LIVES** *solitarily in northern forests; feeds in open ground, water, and marshes.*

The Elk, or Moose as it is known in North America, is a very large, long-legged, almost tailless deer, with a uniform dark grey-brown coat. It has a distinct facial profile, with an inflated end to the long muzzle, overlapping the mouth. Males generally have flattened, spreading antlers, which are usually palmate; although in rare cases they are simply branched.

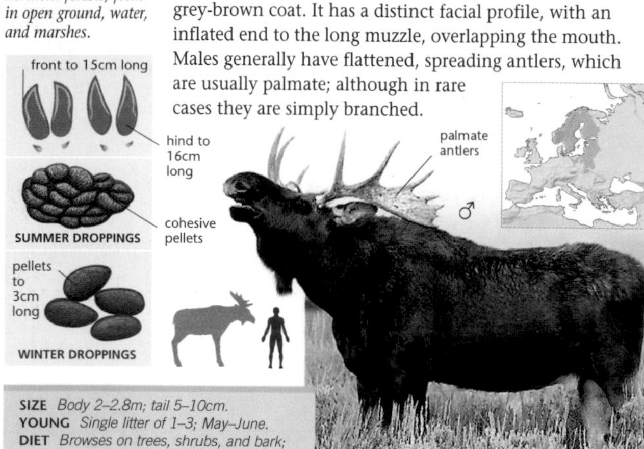


**SIZE** *Body 2–2.8m; tail 5–10cm.*
**YOUNG** *Single litter of 1–3; May–June.*
**DIET** *Browses on trees, shrubs, and bark; also wetland and lake vegetation, especially during the summer.*
**STATUS** *Locally common.*
**SIMILAR SPECIES** *None.*

# Reindeer

*Rangifer tarandus* (Artiodactyla)

**SEEN** *in open montane or Arctic tundra; often moving to open woodland in winter.*

Unique among its family, both male and female Reindeer have antlers, albeit smaller in the female. The antlers are asymmetric, and sometimes palmate at the ends. They have grey-brown fur that turns paler in winter, white underparts, and large feet (that help them to spread their weight on snow). Reindeer are known as Caribou in North America.

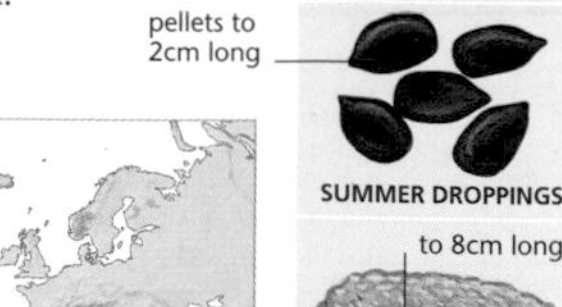


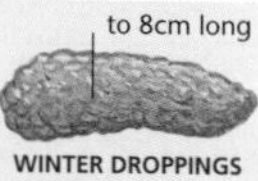


**SIZE** *Body 1.8–2.1m; tail 10–15cm.*
**YOUNG** *Single young; May–June.*
**DIET** *Tundra plants in summer; lichens exposed by hoof scrapings in winter.*
**STATUS** *Wild populations scarce; semi-domesticated form much more common.*
**SIMILAR SPECIES** *None.*

# Red Deer

*Cervus elaphus* (Artiodactyla)

A mature male Red Deer with a full set of antlers is a magnificent sight, each antler bearing five or more points, with two forward points near the base. These antlers and its noisy, bellowing rut behaviour, enable a dominant male to secure a considerable harem of breeding females. All adult animals can be distinguished by their uniform red-brown coat (young fawns are spotted with white), and their buff-coloured rump, without a black outline. These large deer leave many signs of their presence, including mud wallows, large black droppings, and extensive damage to saplings and the bark of larger trees.

**FAVOURS** *open, deciduous woodland, parkland, and riverine marshes; also on open mountains and moorland, though descending into woodland for the winter.*

**NOTE**

*While the antlers of adult males are distinctive, younger animals have less developed antlers, and each antler is itself fully grown only from autumn to spring. There is, therefore, plenty of potential for confusion between deer species, except when adult males are in peak condition.*

♂

branched antlers

large ears

red-brown coat (greyer in winter)

**JUVENILE**

white spots

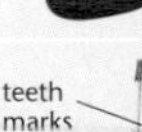

teeth marks 1.6cm wide

**GNAWED BARK**

**SIZE** *Body 1.6–2.6m; tail 10–15cm.*
**YOUNG** *Single young; born May–June.*
**DIET** *Grazes, and browses heather, conifers, birch, and other trees; strips bark.*
**STATUS** *Endangered (Corsican form); otherwise locally common.*
**SIMILAR SPECIES** *Sika Deer (p.118), which is smaller and has a white rump; White-tailed Deer (p.118), which has simpler antlers.*

# Sika Deer

*Cervus nippon* (Artiodactyla)

**FOUND** *in deciduous and coniferous woodland, and adjacent open habitats.*

The Sika Deer can always be distinguished by its black-bordered white rump, and mostly white tail. In summer, adults are lightly spotted, although in winter the coat is greyer and the spots less obvious. Interbreeding between the Sika and Red Deer represents a considerable risk to the genetic integrity of the native Red Deer.

WINTER COAT

**SIZE** *Body 1.2–1.4m; tail 12–15cm.*
**YOUNG** *Single young; May–June.*
**DIET** *Grass, herbs, low shrubs, and trees.*
**STATUS** *Introduced; locally common.*
**SIMILAR SPECIES** *Red Deer (p.117); Fallow Deer (p.119); Spotted Deer* (Axis axis), *which has a white throat but no black on the rump.*

# White-tailed Deer

*Odocoileus virginianus* (Artiodactyla)

**INHABITS** *woodland, both deciduous and coniferous, feeding in open areas and agricultural land.*

Introduced from North America, the White-tailed Deer has a reddish coat in summer; in winter, it is greyer with a dark breast patch. It displays a prominent white rump when running, as it holds its tail up, exposing the white underside of the tail and the rump patch. Adult males have relatively simple, forward-arching antlers.

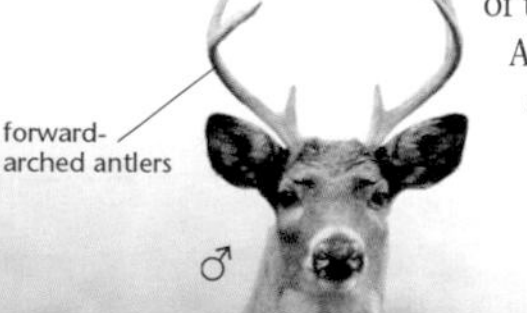


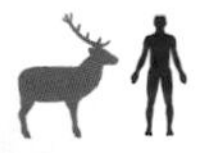

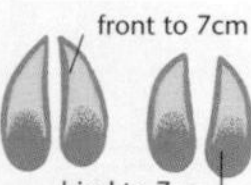


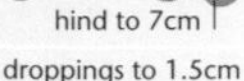

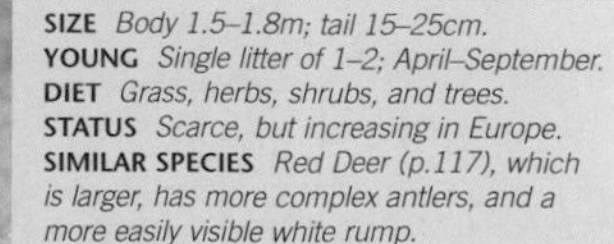

**SIZE** *Body 1.5–1.8m; tail 15–25cm.*
**YOUNG** *Single litter of 1–2; April–September.*
**DIET** *Grass, herbs, shrubs, and trees.*
**STATUS** *Scarce, but increasing in Europe.*
**SIMILAR SPECIES** *Red Deer (p.117), which is larger, has more complex antlers, and a more easily visible white rump.*

# Fallow Deer

*Dama dama* (Artiodactyla)

**FAVOURS** *open woodland and parkland, and adjacent agricultural habitats.*

A familiar and widespread deer, the Fallow Deer is found over much of Europe as a result of introductions and escapes; indeed, many free-ranging herds remain semi-domesticated. The ground colour of a Fallow Deer's coat is usually orange-brown, boldly spotted with white in summer, and greyer and almost unspotted in winter. At all seasons, the white rump with a black border, and long tail with a blackish upper surface, are distinctive, as are the flattened, palmate antlers of an adult male. Look out for trampled rutting rings and a distinct browse line in its natural habitat.

palmate antlers in adult male

antlers with single basal point

white-spotted summer coat

♂

BROWSE LINE

STRIPPED BARK

**NOTE**

*The ground colour of the Fallow Deer ranges from black to white; the paler ones often lack the black border to the rump; the degree to which spots are lost in winter also varies.*

**SIZE** *Body 1.3–1.6m; tail 16–20cm.*
**YOUNG** *Single young; born June–July.*
**DIET** *Grazes in open woodland or on nearby crops; browses saplings and larger trees; acorns and fruit; strips bark in the winter.*
**STATUS** *Native to the Mediterranean; widely introduced and common elsewhere.*
**SIMILAR SPECIES** *Sika Deer (p.118), which has a shorter, white tail.*

# Muntjac

*Muntiacus reevesi* (Artiodactyla)

**OCCURS** *in dense woodland, especially deciduous; feeds in clearings, coppice plots, fields, and gardens.*

A very small deer, the Muntjac is generally dark brown above and white below, with a bushy tail that almost covers its white rump. When alarmed, it runs with the tail raised, revealing its white rump patch. Males have very short antlers, borne on skull projections which remain even after the antlers are shed.

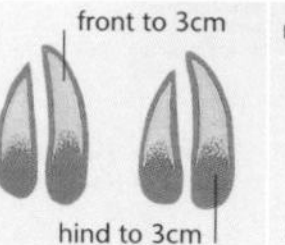


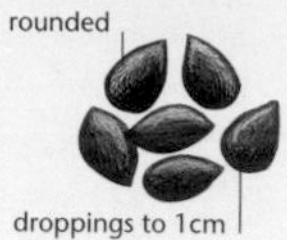


**SIZE** *Body 90–105cm; tail 14–18cm.*
**YOUNG** *One or two litters of 1–2.*
**DIET** *Browses woodland shrubs, Bramble leaves, fruit, acorns, Bracken, and Ivy.*
**STATUS** *Introduced, and locally common.*
**SIMILAR SPECIES** *Chinese Water Deer (below); Roe Deer (p.121).*

# Chinese Water Deer

*Hydropotes inermis* (Artiodactyla)

**FOUND** *in deciduous, especially damp, woodland, marshes, and reedbeds.*

The only European deer never to bear antlers, adult male Chinese Water Deer instead have upper canines elongated into tusks, up to 8cm long, which visibly protrude from the mouth. It is also the only species not to have a distinctly coloured rump patch. Its solitary nature, small size, and largely nocturnal habits make it difficult to see.

droppings to 1.5cm

**SIZE** *Body 80–105cm; tail 4–8cm.*
**YOUNG** *Single litter of 1–2; May–July.*
**DIET** *Grazes grasses, sedges, and crops; browses shrubs and low trees.*
**STATUS** *Introduced; scarce but increasing.*
**SIMILAR SPECIES** *Muntjac (above); Roe Deer (p.121), both of which have antlers.*

# Roe Deer

*Capreolus capreolus* (Artiodactyla)

The most widespread European deer, and the smallest native species, the Roe Deer is a secretive, solitary, and nocturnal woodland inhabitant, although it is increasingly found in suburban habitats. Red-brown above, turning grey in winter, its most distinctive feature is the white rump patch and short white tail. As with most deer, the fawns are spotted with white. The antlers of the male are short, with at most three points each.

**FAVOURS** *woodland, both deciduous and coniferous, with a dense shrub layer, tall Mediterranean scrub, open farmland, reedbeds, and moorland.*

**NOTE**

*Field signs of Roe Deer include frayed bark of trees; it rubs against bark to help remove the "velvet" from its developing antlers, and also to act as a territorial marker for the rut.*

♀

unspotted red-brown coat

white-spotted coat

**FAWN**

small antlers

♂

white upper lip and chin

both prints to 4.5cm

front print splayed

droppings to 1.8cm

frayed bark

**FIELD SIGN**

**SIZE** *Body 1–1.4m; tail 2–3cm.*
**YOUNG** *Single litter of one, occasionally two, May–June.*
**DIET** *Browses woodland-edge shrubs and low trees; also eats crops and autumn berries.*
**STATUS** *Common and increasing.*
**SIMILAR SPECIES** *Muntjac and Chinese Water Deer (p.120), which are both a little smaller and have a different rump pattern.*

# Reptiles

As cold-blooded, largely terrestrial animals, reptiles are most abundant in the warmer parts of the world where they can more easily maintain their temperature within safe limits. Of the more than 8,000 species known to exist today, only about 125 occur in Europe (including the Green Lizard, pictured below), with the greatest abundance and diversity in the Mediterranean zone. This book deals with nearly 90 species, including all the widespread ones. Closely-related species are grouped together, apart from small lizards, which are arranged geographically to make comparison and identification easier.

TURKISH GECKO

WORM-SNAKE

FOUR-LINED SNAKE

SPUR-THIGHED TORTOISE

# Hermann's Tortoise

*Testudo hermanni* (Chelonia)

The most widely distributed European tortoise, Hermann's Tortoise is a common sight in much of the Balkans and on several Mediterranean islands. It has a strongly domed shell, sometimes with greenish tones, overlaid with varying amounts of dark brown; older tortoises tend to be darker, often scarred, and rather lumpy. This species hibernates between October and March–May (depending upon the local temperature) in dense leaf litter or a hole it excavates in soft ground, although it will emerge to bask in warm winter weather. Males differ from females in having a shorter tail and a concave underside shell.

**PREFERS** *hot, dry scrubby habitats, open woodland, cultivated land, and sand dunes; needs access to shade during the heat of the hottest days.*

REAR PLATES

**NOTE**

*Although generally silent, apart from hissing when disturbed, tortoises are often detected by the noises they create, including the rustle as they push through dry vegetation, and the regular knocking of shells during mating.*

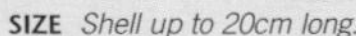

**SIZE** *Shell up to 20cm long.*
**YOUNG** *One or two clutches of up to 12 eggs laid in holes in the ground in early summer; hatch after 2–3 months.*
**DIET** *Herbaceous vegetation and fruit; some carrion and invertebrates.*
**STATUS** *Near-threatened; common in the east.*
**SIMILAR SPECIES** *Spur-thighed Tortoise (p.124), which has distinct spurs and just one tail plate.*

# Spur-thighed Tortoise

*Testudo graeca* (Chelonia)

**OCCUPIES** *hot, dry grassland and scrub, sand dunes, and open woodland, including foothills of mountains.*

This tortoise is usually recognized by its spurs, a single tail plate, coarsely scaled front legs, and a smooth shell profile. However, none of these features is constant, so a combination is needed for certain identification. It is also geographically variable: western tortoises have a paler upper shell, symmetrical black blotches below, and a black and yellow head pattern; in the east they are darker, less regular below, and have a more uniform head colour.

coarse scaling

single tail plate

**SIZE** *Shell up to 30cm long.*
**YOUNG** *Lays 1–4 clutches of eggs; May–July.*
**DIET** *Herbaceous vegetation and fruit.*
**STATUS** *Vulnerable; locally common.*
**SIMILAR SPECIES** *Marginated Tortoise (below), which has weak spurs; Hermann's Tortoise (p.123), which has two tail plates.*

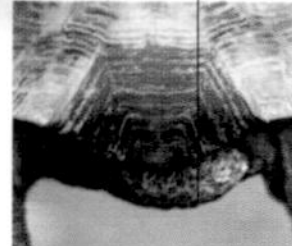

REAR PLATES

# Marginated Tortoise

*Testudo marginata* (Chelonia)

**FOUND** *in dry, scrubby grassland, which is often rocky, and mountain foothills.*

A large, dark tortoise, with a long, distinctly flared shell, an adult Marginated Tortoise is easily recognized; the older they are the darker they get. However, it is possible to confuse juveniles with the Spur-thighed Tortoise. Locality is, perhaps, the best differentiating feature between the two species as they have only a very minor range overlap in northern Greece.

REAR PLATES

weak spurs on thighs

dark brown plates with pale patch

**SIZE** *Shell up to 35cm long.*
**YOUNG** *Up to 15 eggs.*
**DIET** *Vegetation, fruit, snails, and earthworms.*
**STATUS** *Locally common.*
**SIMILAR SPECIES** *Spur-thighed Tortoise (above), which has distinct spurs and at best a weakly flared shell.*

# European Pond Terrapin

*Emys orbicularis* (Chelonia)

The only member of its family to be found naturally outside the New World, the European Pond Terrapin is distinguished from other terrapins of the region by its neck colour and pattern. Its neck, and sometimes legs, are covered in yellowish spots. It has been suggested that its range may have been extended by introductions and that this species is beyond the climatic limit for breeding; however, fossil evidence indicates that it was once found well to the north of its present range. European Pond Terrapins hibernate in soft mud, and also aestivate in similar locations during droughts.

**INHABITS** *still or slow-moving freshwater habitats, usually well vegetated and often with overhanging branches; can tolerate brackish water.*

**NOTE**

*European Pond Terrapins are very wary; they bask on sunny banks, logs, and stones, close to the edge of the water. However, to see them clearly requires a careful, quiet approach.*

**SIZE** *Shell 20–30cm long.*
**YOUNG** *Up to 10 eggs; hatch after 2–4 months or following summer.*
**DIET** *Fish, amphibians, invertebrates, carrion, and vegetable matter.*
**STATUS** *Near-threatened; locally common, but declining in its range.*
**SIMILAR SPECIES** *Spanish and Balkan Terrapins (p.126), which have striped necks; Red-eared Terrapin (p.127), which has red patches behind its eyes.*

# Spanish Terrapin

*Mauremys leprosa* (Chelonia)

**LIVES** *in open bodies of water, including lakes, rivers, and small, shallow, ephemeral pools.*

Considered until recently to be different forms of the Stripe-necked Terrapin, both the Spanish and Balkan Terrapins are restricted to the Iberian and Balkan Peninsulas, respectively. The Spanish Terrapin's shell has a raised keel down the mid-line of the back, and it has pale stripes on the neck and pale eyes. Its long oval shell is broader behind.

long, oval shell

grey-brown shell

**SIZE** *Shell up to 25cm long.*
**YOUNG** *Lays 2–3 clutches of up to 10 eggs.*
**DIET** *Largely carnivorous.*
**STATUS** *Locally common.*
**SIMILAR SPECIES** *Balkan Terrapin (below); European Pond Terrapin (p.125), which lacks neck stripes; Red-eared Terrapin (p.127).*

# Balkan Terrapin

*Mauremys rivulata* (Chelonia)

**FAVOURS** *lowland fresh or brackish water, such as lakes, ponds, and marshes; tolerant of pollution.*

This species has a similar shell to that of the Spanish Terrapin, its western counterpart. It also has stripes on its neck, usually yellow-green, and sometimes an additional stripe down the back of its neck, narrower than those on the side. Its pale undershell is less suffused with dark pigment, which is often restricted to the edges of the plates.

**SIZE** *Shell up to 25cm long.*
**YOUNG** *Lays 2–3 clutches of up to six eggs.*
**DIET** *Small fish, amphibians, invertebrates.*
**STATUS** *Common.*
**SIMILAR SPECIES** *Spanish Terrapin (above); European Pond Terrapin (p.125); Red-eared Terrapin (p.127).*

# Red-eared Terrapin

*Trachemys scripta* (Chelonia)

Introduced from North America, adult Red-eared Terrapins have a uniform dark grey-brown shell. The shell pattern of juveniles is much more distinctive, with complex pale yellowish markings making them much valued as pets. Their feet are webbed and aid the terrapin in swimming. The male is usually smaller than the female with a much longer, thicker tail. Males also have elongated claws that they use in courtship and mating. As they grow, they lose much of their visual appeal and outgrow their aquaria, so they are often discarded into the wild. The distribution as plotted on the map is at best indicative, as Red-eared Terrapins can occur wherever they escape or are released from captivity.

**FREQUENTS** *lakes, ponds, canals, and slow-moving rivers; often close to human habitation, the source of its introduction.*

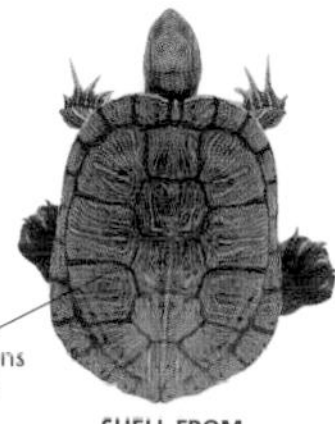


**SHELL FROM ABOVE**

bright red patch behind eyes

pale eyes

clear yellow stripes on neck

**NOTE**

*Unlike many aquatic animals, terrapins are generally tolerant of polluted, cloudy, and brackish water, and thus may be found in the most unpromising sites, including urban centres and tourist areas, where they can be attracted to bread and other scraps.*

**SIZE** *Shell up to 28cm long.*
**YOUNG** *Lays clutches of up to 10 eggs.*
**DIET** *Aquatic invertebrates and plants, fish.*
**STATUS** *Scarce, but increasing.*
**SIMILAR SPECIES** *Native terrapins (pp.125–127); Painted Terrapin* (Pseudemys picta)*, which has a yellow eye-spot.*

# Loggerhead Turtle

*Caretta caretta* (Chelonia)

**LOOK** *for tracks of females visting sandy beaches to breed at night; may be seen by day in shallow coastal waters, or as a tideline stranding.*

Sea turtles are truly oceanic wanderers, rarely approaching land except for breeding. The Loggerhead Turtle is the species most frequently observed in Europe, mostly juveniles that have drifted to the western seaboard on the Gulf Stream, and as a breeding species in the Canary Islands and (locally) in the Mediterranean. The adult has a very large, elongated oval, red-brown shell which is paler yellow-green beneath.

BEACH TRACKS

**SIZE** *Shell up to 1.2m long.*
**YOUNG** *Up to six clutches of up to 200 eggs.*
**DIET** *Marine crustaceans; plant material.*
**STATUS** *Endangered; rare.*
**SIMILAR SPECIES** *Green Turtle* (Chelonia mydas), *which is large and oval; Hawksbill Turtle* (Eretmochelys imbricata).

# Leathery Turtle

*Dermochelys coriacea* (Chelonia)

**INHABITS** *deeper oceanic waters, coming inshore to breed, but not in Europe; seen usually as an accidental stranding.*

The only cool-water sea turtle, the Leathery Turtle is found in the Atlantic, Pacific, and Indian Oceans; it is the most widely distributed reptile in the world. It maintains its body temperature by swimming actively, then conserving heat generated by the muscles with a thick layer of fat. It is very large, and the ridged shell is covered in leathery skin.

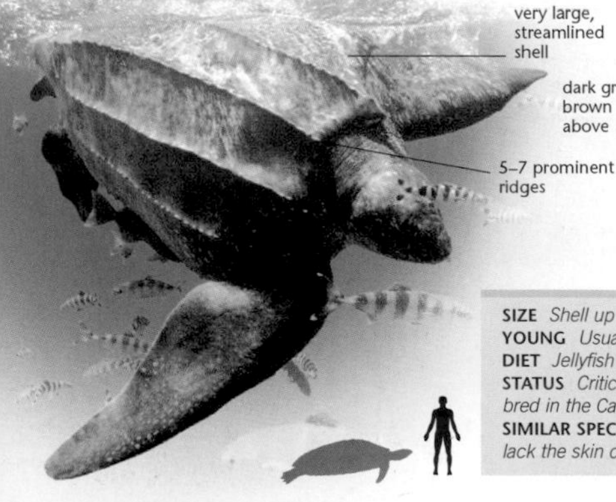


dark grey-brown above

EGG-LAYING FEMALE

**SIZE** *Shell up to 2.9m long.*
**YOUNG** *Usually breeds only in the Tropics.*
**DIET** *Jellyfish and sea squirts.*
**STATUS** *Critically Endangered; rare, but has bred in the Canaries.*
**SIMILAR SPECIES** *None; all other sea turtles lack the skin covering the shell.*

# Agama

*Laudakia stellio* (Squamata)

The only representative of a largely tropical family, the Agama is a robust, flattened, spiny lizard, which is unmistakable in Europe. The skin is grey to sandy brown, although it changes to match its environment, and there is usually a line of pale yellowish, diamond-shaped patches running down the back, which are especially obvious on males. However, there is much variation in colour in different parts of its range and at different temperatures (lighter when warm and darker when cold). The tail often shows several dark bars. It frequently basks in the sun on rocks, walls, and roofs, allowing good views if approached with care.

**FAVOURS** *hot, dry, and sunny habitats, especially rocky hillsides, stone walls, and old buildings, where there are crevices for refuge.*

DARKER FORM

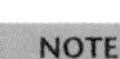

**NOTE**

*Agamas can scale vertical faces very effectively, using their long, sharp claws and rough tail; when they stop, they typically bob their head up and down.*

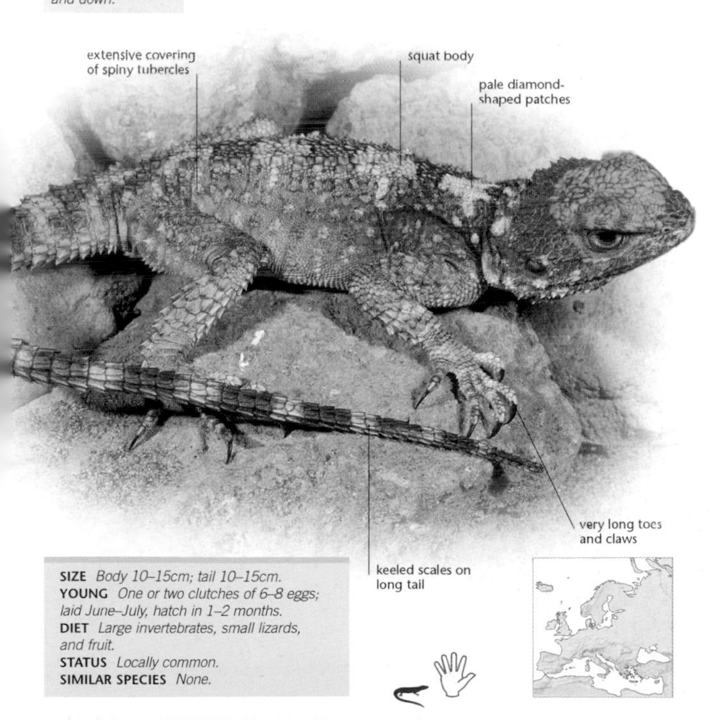


**SIZE** *Body 10–15cm; tail 10–15cm.*
**YOUNG** *One or two clutches of 6–8 eggs; laid June–July, hatch in 1–2 months.*
**DIET** *Large invertebrates, small lizards, and fruit.*
**STATUS** *Locally common.*
**SIMILAR SPECIES** *None.*

# Mediterranean Chameleon

**INHABITS** *bushes and low trees in warm, dry habitats, often in coastal areas; normally descends to the ground only to lay eggs.*

## *Chamaeleo chamaeleon* (Squamata)

The Mediterranean Chameleon is unmistakable over the vast majority of its European range. Its skin colour is determined by the surrounding environment and temperature, and is capable of rapid change. Mostly spending its time climbing bushes, it grasps the branches with its tail and modified feet – the toes of each foot are fused into two gripping claws. It is rather shortlived, typically surviving for about three years, and as a result, numbers are prone to wide fluctuations.

laterally flattened body

prehensile tail

COLOUR FORM

**NOTE**

*Stealthy movement and excellent camouflage enable this chameleon to approach potential prey; the eyes swivel forward to focus on and judge distance to it, and the very long, sticky tongue is unleashed to make the catch.*

broad head

COLOUR FORM

**SIZE** *Body 10–15cm; tail 12–15cm.*
**YOUNG** *Up to 40 eggs laid usually in late summer; hatch the following spring.*
**DIET** *Insects, small lizards, and very young birds.*
**STATUS** *Rare and local.*
**SIMILAR SPECIES** *African Chameleon* (C. africanus), *a larger species without head-flaps, which is found locally in S. Greece.*

# Moorish Gecko

*Tarentola mauritanica* (Squamata)

Significantly larger and more robust than other European geckos, the Moorish Gecko has adhesive pads which extend along the full length of the toes, enabling it to climb vertical faces and overhangs with ease. Its scales are conspicuously keeled, giving it a spiny appearance, especially on the tail. Its grey-brown colour is variable, and capable of changing rapidly to suit its environment. Active in the day and at night, this gecko assumes a light and dark appearance accordingly. Only Moorish Geckos towards the north of their range enter hibernation, in rock crevices or holes in the ground.

**FAVOURS** *dry lowland and coastal areas, usually on rocks, cliffs, and stone walls that are warmed by the sun.*

vertical pupil

**NOTE**

*Other* Tarentola *species on the Atlantic islands are best distinguished by distribution:* T. angustimentalis *(Fuerteventura, Lanzarote),* T. delalandii *(Tenerife, La Palma),* T. boettgeri *(Gran Canaria, El Hierro, Selvages islands), and* T. gomerensis *(La Gomera).*

claws only on the third and fourth toes

PALE FORM

large eyes

broad, flat head

strongly projecting scales

**SIZE** *Body up to 8cm; tail 10cm.*
**YOUNG** *One to three clutches of 1–2 eggs, laid in early summer in rock crevices; hatch after 3–4 months.*
**DIET** *Insects and spiders, often captured around artificial light.*
**STATUS** *Common.*
**SIMILAR SPECIES** *Other mainland geckos (pp.132–133), which are similar but smaller.*

# Turkish Gecko

*Hemidactylus turcicus* (Squamata)

**FOUND** *among ruins, stone walls, cliffs, and on dry hillsides, often entering buildings, even in towns, in search of food.*

*Hemidactylus*, meaning "half-fingered", refers to one of the most distinctive features of the Turkish Gecko – the adhesive pads on its feet do not extend to the end of the toes. This small gecko is covered in projecting scales which give it a spiny appearance, especially on its tail. Its colour ranges from whitish and translucent to dark brown, often with reddish blotches and a conspicuously barred lower tail. Turkish Geckos are active mainly at night, when they are often to be heard producing a variety of noises, from clicks to mewing – geckos are some of the few reptiles with a true voice.

**NOTE**

*The toe pads of geckos are covered in microscopic, adhesive hairs, allowing them to climb agilely up any surface, including the smoothest of walls, and even cling to ceilings in search of prey.*

DARK FORM

**SIZE** *Body up to 6cm; tail 6cm.*
**YOUNG** *Two or three clutches of one or two eggs, laid in crevices or burrows; hatch after 1–3 months.*
**DIET** *Insects, especially attracted to artificial lights.*
**STATUS** *Common.*
**SIMILAR SPECIES** *Other geckos (pp.131, 133), which are not as pale and have less distinctive toes.*

# European Leaf-toed Gecko

*Euleptes europaea* (Squamata)

The smallest of European geckos, a lack of projecting scales gives this gecko a smooth-skinned appearance. The adhesive pads on this species are restricted to the tips of the toes. Typically, it is a grey-brown colour and the lower half of the tail is often swollen, especially if the tail has regrown after falling off. This is the only European gecko that does not always hibernate.

**FAVOURS** *rocks, walls, and dead trees, shaded from the midday sun; often on the outside of buildings but does not freely enter them.*

often swollen tail

smooth skin

dark transverse colour bands

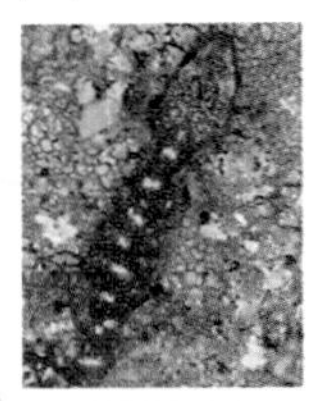

CAMOUFLAGE

**SIZE** *Body up to 4cm; tail 4cm.*
**YOUNG** *Up to three clutches of one or two eggs; hatch after 2–4 months.*
**DIET** *Insects and other invertebrates.*
**STATUS** *Locally common.*
**SIMILAR SPECIES** *Young of other geckos could be confused with it.*

# Kotschy's Gecko

*Mediodactylus kotschyi* (Squamata)

Slender, with more of the appearance of a rock lizard, Kotschy's Gecko lacks adhesive toe pads, relying on its very long toes and sharp claws for its climbing abilities: it is not able to climb as far or as well as its relatives. Although largely nocturnal, it does occasionally emerge in daylight, and in strong sunshine, its grey-brown body turns almost black.

**PREFERS** *dry, stony ground, especially granite or volcanic rock features, but will occupy walls and the outside of buildings.*

PALE FORM

**SIZE** *Body up to 5cm; tail 5cm.*
**YOUNG** *Clutches of one or two eggs; hatch after 3–5 months.*
**DIET** *Insects and other invertebrates.*
**STATUS** *Locally common.*
**SIMILAR SPECIES** *None, especially if the toes are clearly seen.*

# Large Psammodromus

*Psammodromus algirus* (Squamata)

**FOUND** *in a range of hot, dry lowland habitats, including open woodland and scrub, cultivated areas, stone walls, and even prickly-pear hedges.*

Also known as Algerian Sand Racer, this shy, fast-moving lizard spends much of its time hunting in scrub and low trees, but its speed and agility are most apparent when it scurries between bushes. Hibernation takes place in rock crevices, and on emergence males assume their breeding colours, including an orange, red, or yellow throat and chest (white at other times), and blue spots above the front legs. Immature animals usually have an orange tail.

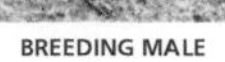


BREEDING MALE

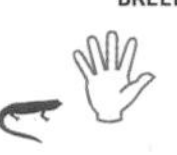

**SIZE** *Body 6–9cm; tail to 20cm.*
**YOUNG** *Two or three clutches of 8–11 eggs; laid April–May; hatch after 3–4 months.*
**DIET** *Insects, spiders, small lizards, and fruit.*
**STATUS** *Common.*
**SIMILAR SPECIES** *Spanish Psammodromus (below), which is much smaller.*

# Spanish Psammodromus

*Psammodromus hispanicus* (Squamata)

**REQUIRES** *hot, dry, and sunny habitats, with shrubs and low trees to provide cover for hunting.*

With overlapping, pointed, keeled scales, the Spanish Psammodromus (or Spanish Sand Racer) is clearly related to the Large Psammodromus, and is a similarly fast-moving species. It has pale spots or streaks forming lines, often broken, down its brown back. In the breeding season, the flank stripes of the male may become bright yellow, and the white underparts flushed with red.

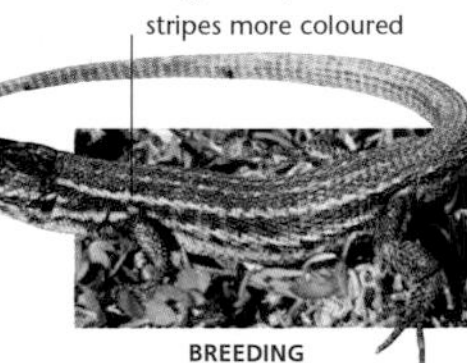


BREEDING

**SIZE** *Body up to 5cm; tail 10–12cm.*
**YOUNG** *Up to six eggs; laid in June; hatch in August.*
**DIET** *Insects and other invertebrates.*
**STATUS** *Common.*
**SIMILAR SPECIES** *Large Psammodromus (above), which is much larger.*

# Spiny-footed Lizard

*Acanthodactylus erythrurus* (Squamata)

The only European representative of a mainly African, desert-dwelling group, which have fringes of scales on their toes to enable them to move swiftly over loose sand, the Spiny-footed Lizard is typically found in open, sandy areas, over which it runs far and fast. When it stops, it usually adopts a "head-up" alert posture. The adult Spiny-footed Lizard is variable in colour, but usually shows pale stripes down the back, separated with broken dark lines. The markings on the legs are a mixture of dark and pale spots, while the underparts of the lizard are whitish. In contrast, young Spiny-footed Lizards have a very distinctive, unique, stripy colour pattern.

**OCCUPIES** *hot and dry sandy and stony areas, usually with sparse scrub, but may be found on completely bare upper beaches.*

JUVENILE

**NOTE**

*Juveniles have a distinct colour and pattern: the body is longitudinally striped black and white, and the tail is orange-red, which sometimes extends to the thighs.*

**SIZE** *Body up to 8cm; tail up to 15cm.*
**YOUNG** *One or two clutches of up to eight eggs; hatch in about two months.*
**DIET** *Insects, especially grasshoppers, other ground-dwelling invertebrates, fruit, and seeds.*
**STATUS** *Locally common.*
**SIMILAR SPECIES** *None.*

# Greek Algyroides

*Algyroides moreoticus* (Squamata)

**FOUND** *on damp, shady, north-facing rocks and slopes, in open woodland, and on stone walls.*

A small, rather inconspicuous and secretive lizard, the Greek Algyroides prefers a more shaded habitat than many other species. The female is dull in colour, with pale brown underparts, although the lack of pigmentation allows the strongly keeled scales on the back and flanks to be clearly seen. The male, on the other hand, is more distinctive, with yellow to reddish underparts.

**SIZE** *Body up to 5cm; tail up to 12cm.*
**YOUNG** *Lays clutches of two eggs.*
**DIET** *Insects and other invertebrates.*
**STATUS** *Very locally common in its range.*
**SIMILAR SPECIES** *Other Algyroides (below and p.137), which overlap very little in their distribution.*

# Dalmatian Algyroides

*Algyroides nigropunctata* (Squamata)

**FAVOURS** *coastal, hot, dry areas, with access to shade when needed, including open scrub and stone walls.*

The Dalmatian Algyroides has large scales on its back, which give it a rough-skinned appearance. It is very agile, using its very long toes to give it a grip on rock surfaces. Although diurnal, it hides from the sun in the hottest periods, and hibernates in rock crevices. Breeding males have blue-green throats and orange bellies.

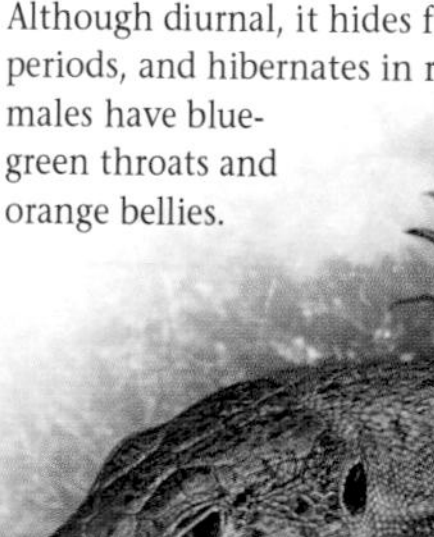

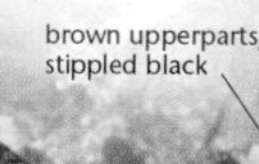


**SIZE** *Body up to 7cm; tail up to 14cm.*
**YOUNG** *One or two broods of 2–3 eggs.*
**DIET** *Insects and other invertebrates.*
**STATUS** *Locally common, though very secretive.*
**SIMILAR SPECIES** *Greek Algyroides (above); Common Wall Lizard (p.143).*

# Pygmy Algyroides

*Algyroides fitzingeri* (Squamata)

Its small range, restricted to Corsica and Sardinia, prevents the Pygmy Algyroides from being confused with other lizards. Its inconspicuous colour and small size make it difficult to observe, despite its range of habitats, from sea-level to mountain slopes. As with most others in its genus, the overlapping scales are characteristic. It has a white chin, yellow belly, and stout tail.

**PREFERS** *semi-shaded areas, often close to water, including open woodland, scrub, and mountain slopes.*

large, pointed, strongly keeled scales

dark brown, sometimes streaked or spotted

flattened body and head

**SIZE** *Body up to 4cm; tail up to 8cm.*
**YOUNG** *Clutches of 2–4 eggs, hatch after 2–3 months.*
**DIET** *Insects and other invertebrates.*
**STATUS** *Locally common, but inconspicuous.*
**SIMILAR SPECIES** *None within its range.*

# Spanish Algyroides

*Algyroides marchi* (Squamata)

Discovered as recently as 1958, the Spanish Algyroides is a small, flattened, and rather insubstantial lizard. Its scales are only weakly keeled, and those on the back are very much larger than the flank scales. Usually a chestnut brown colour, there is sometimes a darker central streak down the back, and the pale underparts turn yellow in the breeding season. In some areas, males develop bright blue throats.

**INHABITS** *woodland and other shady habitats, especially close to streams; it climbs well and is found in bushes and trees.*

small scales on flanks

overlapping back scales

collar below neck

**SIZE** *Body up to 5cm; tail up to 10cm.*
**YOUNG** *Clutches of 2–5 eggs, hatch in 1–2 months.*
**DIET** *Insects and other invertebrates.*
**STATUS** *Vulnerable; generally scarce.*
**SIMILAR SPECIES** *Other Algyroides which have markedly different distributions.*

# Ocellated Lizard

## *Timon lepidus* (Squamata)

**LIVES IN** *dry lowland habitats, including cultivated areas, scrub, and open woodland, with bushes and holes in which to take refuge.*

The largest mainland member of its family, the Ocellated Lizard is an impressive, robust green lizard, usually with bright blue spotting on its flanks. The male has an especially large head, and powerful jaws, which it uses both to catch prey and in self-defence – it will bite hard if handled. A diurnal hunter, it hibernates between October and March in burrows and rock crevices. There is some geographic variation in colour, and it is suggested that some of these variants may merit recognition as separate species.

JUVENILE

**NOTE**

*The ocelli, or eye-spots, (white with a black border) after which this species is named are especially noticeable on the juvenile Ocellated Lizard.*

green body, stippled black

blue-spotted flanks

large, broad head

powerful limbs

♂

**SIZE** *Body up to 20cm; tail 30–40cm.*
**YOUNG** *One or two clutches of up to 20 eggs; hatch after 2–4 months.*
**DIET** *Beetles, other large insects, birds' eggs, chicks; other reptiles and amphibians; occasionally young rabbits; fruit and other plant matter.*
**STATUS** *Locally common.*
**SIMILAR SPECIES** *Schreiber's Green Lizard (p.139) and Green Lizard (p.140), which are both smaller and lack the blue flank spots.*

# Schreiber's Green Lizard

*Lacerta schreiberi* (Squamata)

This moderately large lizard has a long, thin tail, up to twice the body length. Males are green above, with black spots; they also have black spotting on the yellow belly, and, when breeding, a blue throat. Females are brown, but also have extensive black spotting, often organized into stripes. Juveniles resemble those of the Ocellated Lizard, although the ocelli are elongate and found mostly on the flanks.

**INHABITS** *open rocky and sandy areas, especially on south-facing slopes.*

♂

black spotted green body

yellowish, black-spotted, belly

JUVENILE

♀

brown, spotted with black

**SIZE** *Body up to 12cm; tail 12–20cm.*
**YOUNG** *Clutches of 10–18 eggs.*
**DIET** *Insects, reptiles, young birds, and fruit.*
**STATUS** *Near-threatened; locally common.*
**SIMILAR SPECIES** *Ocellated Lizard (p.138), which is larger with blue flank spots; Green Lizard (p.140), which lacks black blotches.*

# Balkan Green Lizard

*Lacerta trilineata* (Squamata)

This lizard has a long tail, up to twice the length of its body. The green body of the male is covered in fine black stippling, and the paler female usually shows yellow-green flank patches and back stripes. Juveniles usually show three (or five) obvious white stripes down the back, giving rise to its scientific name; some more uniform individuals have pale flank spots without a black border.

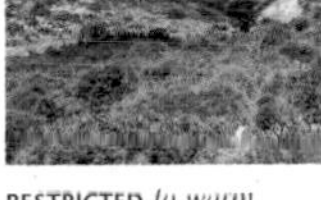

**RESTRICTED** *to warm lowland habitats, especially scrubby slopes, sand dunes, stone walls, and ruins.*

JUVENILE

odd number of stripes

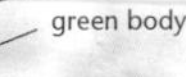

♂

**SIZE** *Body up to 16cm; tail 20–30cm.*
**YOUNG** *One or two clutches of 9–18.*
**DIET** *Insects, small vertebrates, and fruit.*
**STATUS** *Common.*
**SIMILAR SPECIES** *Green Lizard (p.140), which is smaller, with a blue throat; the juvenile has an even number of stripes.*

# Green Lizard

*Lacerta viridis* (Squamata)

**OCCUPIES** *dry habitats, including open woodland, heathland, and cultivated areas, up to sub-alpine levels; also riverbanks and damp areas, especially in the south.*

Over much of Europe, the Green Lizard is the largest lizard to be found. It is primarily green, though the male is finely stippled with black, and has a blue throat, especially noticeable in the breeding season. Females tend to be a little browner, with two or four indistinct pale stripes, and sometimes dark blotches. Hibernation takes place only in the northern part of their range. Green Lizards frequently bask in open areas, though they retreat into shade if it is too hot. Riverbanks are often favoured sites, as they can take refuge from disturbance by diving into the water, and burying themselves in soft mud.

**NOTE**

*In areas where the range of Green and Balkan Green Lizards overlap, look out for stripy juveniles, which are easier to tell apart: Balkan Green has 3 or 5 stripes, whereas Green has 2 or 4 – it lacks a stripe down the spine.*

**SIZE** *Body up to 13cm; tail up to 30cm.*
**YOUNG** *4–20 eggs laid in June; hatch after 2–4 months.*
**DIET** *Insects, worms, reptiles, rodents, birds' eggs, and nestlings; fruit.*
**STATUS** *Common, especially in the south of its range.*
**SIMILAR SPECIES** *Ocellated Lizard (p.138), which has blue flank spots; Schreiber's Green Lizard (p.139), which has larger black blotches; Balkan Green Lizard (p.139), which is larger; Sand Lizard (p.141).*

# Sand Lizard

*Lacerta agilis* (Squamata)

The Sand Lizard is powerfully built, with a short head and legs, and a stocky body. It is a ground-dweller, with poor climbing abilities. One of the most variable and beautifully marked lizards, it has a band of narrow scales running down its back; this band is often a different colour to the rest of the body, but only rarely green. Otherwise, it is difficult to generalize about the colour: green flanks and chestnut brown backs are common, some have pale stripes, sometimes including a narrow white line down the spine, and the entire upperpart can be blotched with black patches or, on the flanks, with white-centred ocelli.

**FOUND** *in dry grassy areas, hedgebanks, heathland, and woodland margins, and into montane habitats in the south.*

blunt snout

eye spots on flanks

♂

flanks often green

JUVENILE

band of narrow scales

NOTE

*After emergence from hibernation, there is intense rivalry between males; with arched backs, puffed-out necks, and open mouths, they charge at each other and grapple violently to secure access to a ripe female.*

**SIZE** *Body up to 9cm; tail up to 15cm.*
**YOUNG** *One or two clutches of up to 14 eggs.*
**DIET** *Insects, slugs, other invertebrates; fruit, flowers; small lizards, including the young of their own species.*
**STATUS** *Locally common.*
**SIMILAR SPECIES** *Green Lizard (p.140), which is larger and more uniformly coloured; Viviparous Lizard (p.142), which is less robust.*

# Viviparous Lizard

*Zootoca vivipara* (Squamata)

**OCCUPIES** *a wide range of habitats, lowland to montane, from open, dry sand dunes to damp, shady woodland; hibernates under logs or stones.*

Sometimes known as the Common Lizard, the Viviparous Lizard is indeed the commonest species over much of Europe, apart from the Mediterranean region. It is also the most northerly occurring species; its range extends into the Arctic zone. Unlike some similar species, its body is not flattened, except when basking. The ground colour varies from grey-brown to reddish and olive green; there are variable stripes down the back, sometimes with black blotches or ocelli, often organized into rows down the back or flanks.

coarse, keeled scales on back

thick neck and poorly differentiated head

BLACK FORM

### NOTE

*The female lizard retains the eggs internally as the young develop, until they are born live. This is an adaptation to its northerly distribution, allowing her to give birth to the young in the warmest places.*

white throat (may be blue in a breeding male)

yellow-orange below

UNDERSIDE

serrated collar edge

dark stripe down back

variable black blotches

**SIZE** *Body up to 6.5cm; tail 8–10cm.*
**YOUNG** *Up to 10 live young; born in damp areas, June–September.*
**DIET** *Insects, spiders, snails, earthworms, and other invertebrates.*
**STATUS** *Common.*
**SIMILAR SPECIES** *Sand Lizard (p.141), which is larger and green; Common Wall Lizard (p.143), which has longer legs and a prominent head; Meadow Lizard (p.152), which has more contrasting stripes.*

# Common Wall Lizard

*Podarcis muralis* (Squamata)

Typically of its group, the Common Wall Lizard has a slender, flattened body, long legs, and a long pointed snout, while the head is clearly wider than its body. Its scales are slightly keeled, giving a roughened appearance. Variable black spotting on its back sometimes form a reticulated pattern. Its underparts are yellowish or cream, turning orange in a breeding male. It is a climbing species, although rather less agile than some of its relatives. Where it is found with more specialized climbers, Common Wall Lizards usually occupy the less precipitous faces.

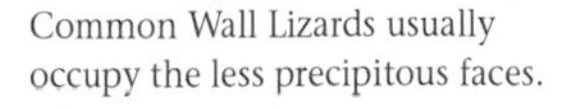

**OCCUPIES** *rocks and cliffs, stone walls and the outside of houses, often foraging in grassy places and open woodland; largely montane in the south.*

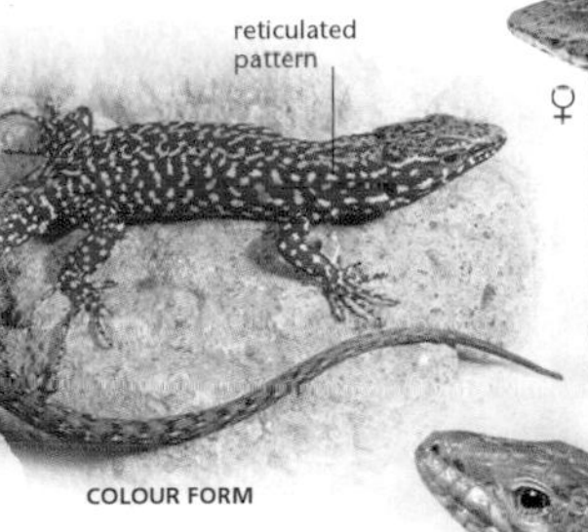


COLOUR FORM

**NOTE**

*Wall lizards are very difficult to tell apart as most are variable. Getting to know the Common Wall Lizard, the most widespread, helps, but often range is the best clue to identification.*

**SIZE** *Body up to 7.5 cm; tail up to 16cm.*
**YOUNG** *Lays up to three clutches of up to 10 eggs, February–June.*
**DIET** *Insects, spiders, crustaceans, fruit, young lizards.*
**STATUS** *Common.*
**SIMILAR SPECIES** *Dalmatian Algyroides (p.136); Viviparous Lizard (p.142); Iberian (p.145), Bocage's (p.146), and Italian Wall Lizard (p.149); Hovarth's Wall Lizard (p.150), which has a spotted throat.*

# Iberian Rock Lizard

*Iberolacerta monticola* (Squamata)

**OCCUPIES** *montane habitats, largely above 1,100m, at or around the tree-line on screes and boulders.*

Generally a high mountain species, the Iberian Rock Lizard descends to rocky coastal areas in Galicia. It is rather small, but robust, and has a long tail. Variable in colour, it may often have strikingly green underparts. Many males are greenish above, heavily spotted with black, while females are yellow-green below with brown stripes above. Juveniles are distinctive, with a bright blue tail.

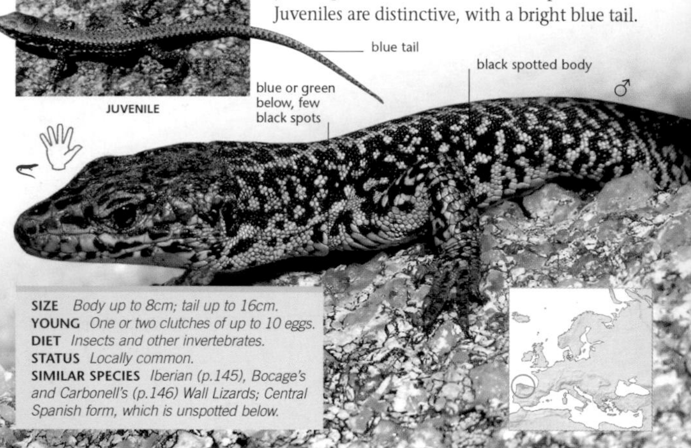


**SIZE** *Body up to 8cm; tail up to 16cm.*
**YOUNG** *One or two clutches of up to 10 eggs.*
**DIET** *Insects and other invertebrates.*
**STATUS** *Locally common.*
**SIMILAR SPECIES** *Iberian (p.145), Bocage's and Carbonell's (p.146) Wall Lizards; Central Spanish form, which is unspotted below.*

# Pyrenean Rock Lizard

*Iberolacerta bonnali* (Squamata)

**FOUND** *only in high mountain areas, around screes, rocks, and meadows, often close to water and to crevices for shelter.*

Restricted to a small area of the west-central Pyrenees, at altitudes of between 1,700m and 3,000m, the Pyrenean Rock Lizard was until recently treated as a subspecies of the Iberian Rock Lizard. However, genetic analysis indicates that they (and two other geographic forms) should be considered full species, despite the relatively minor colour and pattern differences. All have extended hibernation periods, dictated by the prolonged snow cover at high altitude.

almost uniform grey-brown above

white throat

obscure pale lines

dark flanks

**DARK FORM**

**SIZE** *Body up to 6cm; tail up to 12cm.*
**YOUNG** *Little known; single clutch of eggs.*
**DIET** *Insects and other invertebrates.*
**STATUS** *Vulnerable; locally common.*
**SIMILAR SPECIES** *Aurelio's Rock Lizard (I. aurelioi) and Aran Rock Lizard (I. aranica) which are stripier and live further east.*

# Iberian Wall Lizard

*Podarcis hispanicus* (Squamata)

**FAVOURS** *cliffs, rocks, stone walls, and open woodland, especially in the lowlands; it tends to avoid human habitation, although it sometimes occupies old buildings.*

Occuring over much of the Iberian Peninsula and taking the place of the Common Wall Lizard, the Iberian Wall Lizard is a delicate species with a flattened body, which displays considerable variation in colour and pattern. Both sexes, but especially females, are usually striped, the stripe down the spine being less obvious than the lateral stripes. Juveniles sometimes show a blue tail, and may therefore be confused with young Iberian Rock Lizards. The high degree of variability suggests that the Iberian Wall Lizard may in fact constitute a group of closely related species. The underparts are generally less variable, and unspotted or nearly so; those with green underparts cannot be mistaken for any other species.

DARK SPOTTED FORM

**NOTE**

*An agile and skilful climber on cliffs and rocks, and into trees, the Iberian Wall Lizard is replaced by other species in less precipitous habitats, such as screes.*

**SIZE** *Body up to 6cm; tail up to 12cm.*
**YOUNG** *Up to five eggs; hatch after about two months.*
**DIET** *Insects and other invertebrates.*
**STATUS** *Common.*
**SIMILAR SPECIES** *Common Wall Lizard (p.143), which has a stronger central stripe; Iberian Rock Lizard (p.144), which is usually green below; Bocage's and Carbonell's Wall Lizards (p.146).*

# Bocage's Wall Lizard

*Podarcis bocagei* (Squamata)

**OCCUPIES** *open deciduous woodland, scrub, and heathland in cool, damp regions.*

More robust and less flattened than the Iberian Wall Lizard with which its distribution overlaps, Bocage's Wall Lizard is a rather variable species in different parts of its range. Females are dark brown with moderately prominent stripes, while males are often green, with pale green stripes on either side of the back. The underparts are generally yellow or orange, and heavily spotted with black. A poor climber, most of its foraging is done on the ground.

**SIZE** *Body up to 7cm; tail up to 14cm.*
**YOUNG** *Three or four clutches of 2–9 eggs.*
**DIET** *Insects and other invertebrates.*
**STATUS** *Locally common.*
**SIMILAR SPECIES** *Carbonell's (below) and Common Wall Lizards (p.143); Iberian Rock Lizard (p.144); Iberian Wall Lizard (p.145).*

# Carbonell's Wall Lizard

*Podarcis carbonelli* (Squamata)

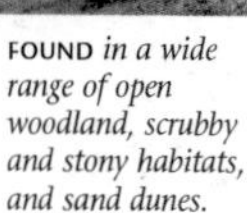

**FOUND** *in a wide range of open woodland, scrubby and stony habitats, and sand dunes.*

Rather similar in appearance to Bocage's Wall Lizard, Carbonell's Wall Lizard overlaps in distribution with it only in a restricted area of the Douro Valley. It is a poor climber, and spends most of its time on the ground. Usually brown above, with greenish flanks, the belly is normally whitish, although there are often blue scales around the edge.

brown above

♂

blue scales

black-spotted back and flanks

**SIZE** *Body up to 6.5cm; tail up to 14cm.*
**YOUNG** *Up to three clutches of 1–4 eggs.*
**DIET** *Insects, snails, and other invertebrates.*
**STATUS** *Locally common.*
**SIMILAR SPECIES** *Bocage's (above) and Common Wall Lizards (p.143); Iberian Rock (p.144) and Wall Lizards (p.145).*

# Lilford's Wall Lizard

*Podarcis lilfordi* (Squamata)

Inhabiting barren Balearic islets where there are few insects, Lilford's Wall Lizard is a robust lizard which has adopted a primarily herbivorous diet, and has an elongated intestine to cope with this. Some native plants even depend upon it for pollination. Green or brown above, it is white, yellow, or pink below. Once common on the main islands, it has largely been eliminated there through the introduction of predators.

**PREFERS** *barren, windswept, salt-affected rocky islets, where more than 20 subspecies exist.*

**STRIPED FORM**

**SIZE** *Body up to 8cm; tail up to 14cm.*
**YOUNG** *Up to three clutches of 1–4 eggs.*
**DIET** *Primarily herbivorous.*
**STATUS** *Vulnerable; locally very common.*
**SIMILAR SPECIES** *Ibiza Wall Lizard (below); Italian Wall Lizard (p.149); Moroccan Wall Lizard* (Teira perspicillata) *is reticulated.*

# Ibiza Wall Lizard

*Podarcis pityusensis* (Squamata)

Populations of Ibizan Wall Lizards on different islands and outcrops are very different, and there are 30 recognizable subspecies. On Ibiza, it is normally robust, with coarse, keeled scales, and is brown or green with broken dark stripes either side of the back, while on Formentera, it is slender, and bright green above. Isolated islet populations have strongly developed blue or orange flanks.

**FAVOURS** *bare rock, with crevices for shelter, stone walls, cultivated fields, open woodland, ruins, and gardens.*

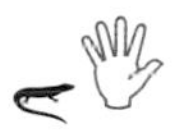

**SIZE** *Body up to 8cm; tail up to 16cm.*
**YOUNG** *Clutches of up to four eggs.*
**DIET** *Invertebrates, especially ants; fruit and seeds.*
**STATUS** *Vulnerable; locally very common.*
**SIMILAR SPECIES** *Lilford's Wall Lizard (above), which is less coarsely scaled.*

# Bedriaga's Rock Lizard

*Archaeolacerta bedriagae* (Squamata)

**SEEN** *climbing bare rocks, cliffs, walls, and old buildings, largely in montane areas, often above the tree-line.*

Found especially on Corsica, with just a few localities on Sardinia, Bedriaga's Rock Lizard is a medium-sized lacertid, with flat, unkeeled scales on its back. This feature distinguishes it from wall lizards occurring in the same areas. It is normally green or grey in colour. Typical of habitual climbing species, it has a strongly flattened body, allowing it to take refuge in small cracks and crevices.

BROWN COLOUR FORM

**SIZE** *Body up to 8cm; tail up to 15cm.*
**YOUNG** *Clutches of three to six eggs.*
**DIET** *Insects and other invertebrates.*
**STATUS** *Locally common.*
**SIMILAR SPECIES** *Tyrrhenian Wall Lizard (below), which has keeled scales on the back; Italian Wall Lizard (p.149).*

# Tyrrhenian Wall Lizard

*Podarcis tiliguerta* (Squamata)

**FOUND** *in lowland fields, scrub, and open woodland, often close to human activity; lives in stone walls.*

The Tyrrhenian Wall Lizard is often seen on the ground as it is a poor climber. Its body is noticeably less flattened than the rock-dwelling Bedriaga's Rock Lizard, which occurs on the same islands but at higher altitudes. Males are generally green, with a dark reticulated pattern, often with blue flank spots; females are brown and stripy, but both have weakly keeled scales on the back, between the legs.

blue flank spots

DARKER MALE

**SIZE** *Body up to 6.5cm; tail up to 12cm.*
**YOUNG** *Clutches of 6–12 eggs.*
**DIET** *Insects and other invertebrates.*
**STATUS** *Locally very common.*
**SIMILAR SPECIES** *Bedriaga's Rock Lizard (above), which has unkeeled scales.*

# Italian Wall Lizard

*Podarcis siculus* (Squamata)

**FREQUENTS** *lowland habitats ranging from meadows and dunes, to vineyards and ruins; very tolerant of humans; seen in town gardens and parks.*

Extreme variability is the cause of much confusion when trying to identify the Italian Wall Lizard: no less than 40 recognizable subspecies have been described over its extensive range. But its shape is more or less constant – it is a robust lizard, with a long, flattened head, and a finely tapering tail. The upperparts are often green, and usually distinctly striped (especially females); males sometimes show dark reticulations. Their underparts are a constant whitish grey, and almost always unspotted. These lizards hibernate, at least in the northern parts of their range, although they frequently emerge to bask on warm winter days.

**NOTE**

*The Italian Wall Lizard has a very wide distribution, as far as north Africa and the USA; this is likely to be due to accidental transport on board ships, a consequence of its lack of fear of humans.*

**SIZE** *Body up to 9cm; tail up to 18cm.*
**YOUNG** *Two to five clutches of up to eight eggs.*
**DIET** *Insects, spiders, slugs, worms, and other invertebrates; fruit.*
**STATUS** *Common.*
**SIMILAR SPECIES** *Common Wall Lizard (p.143); Lilford's Wall Lizard (p.147); Bedriaga's Rock Lizard (p.148); Sicilian Wall Lizard (p.150); Aeolian Wall Lizard* (P. raffonei), *which is brown with a spotted throat.*

# Horvath's Wall Lizard

*Iberolacerta horvathi* (Squamata)

**OCCURS** *only in rocky habitats in mountains, in humid forest areas, up to and around the tree line, and in gorges lower down.*

A flattened, agile climber with smooth, shiny scales, the dull brown Horvath's Wall Lizard is a similar colour to the Common Wall Lizard, but is usually found on steeper rocks and cliffs. Horvath's also has beige eyes and a white unspotted throat, differentiating it from the red eyes and spotted throat of the Common Wall Lizard. It has smooth scales on the back and its belly is pale yellow.

**SIZE** *Body up to 6.5cm; tail up to 12cm.*
**YOUNG** *Lays 3–5 eggs.*
**DIET** *Insects and other invertebrates.*
**STATUS** *Locally common.*
**SIMILAR SPECIES** *Common Wall Lizard (p.143); Mosor Rock Lizard (p.151); Meadow Lizard (p.152).*

# Sicilian Wall Lizard

*Podarcis P. waglerianus* (Squamata)

**LIVES** *in a variety of inland habitats, including woodland edges, scrub, and gardens, sometimes in rather damp areas.*

Generally a ground-dweller, as it climbs only poorly, the Sicilian Wall Lizard usually favours inland areas, in contrast to the Italian Wall Lizard, which is dominant in the coastal belt of Sicily. Dark spots, if present on the often greenish body, are arranged in rows rather than reticulated patterns, and one or two pale stripes on each side may be present, especially in females.

**SIZE** *Body up to 7.5cm; tail up to 12cm.*
**YOUNG** *Lays up to six eggs.*
**DIET** *Insects, invertebrates, fruit, and seeds.*
**STATUS** *Common, especially inland.*
**SIMILAR SPECIES** *Sicilian form of Italian Wall Lizard (p.149), which is usually reticulated, with a pale belly.*

# Dalmatian Wall Lizard

*Podarcis melisellensis* (Squamata)

Rather variable, cylindrical, and stocky, the Dalmatian Wall Lizard is either green or brown, and may be uniform or striped. Unstriped individuals may be green above, with contrasting brown flanks; striped ones often have a broken black central back stripe, with a series of larger dark patches, sometimes including white flecks, down the flanks. The belly is deep orange in breeding males.

**INHABITS** *open, dry areas of scrub, stony hillsides, pastures, open woodland, and stone walls.*

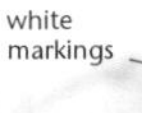

white markings

♂

broken black stripe

♀

**SIZE** *Body up to 6.5cm; tail up to 12cm.*
**YOUNG** *Lays a single clutch of 4–8 eggs.*
**DIET** *Insects and other invertebrates.*
**STATUS** *Common.*
**SIMILAR SPECIES** *Balkan Wall Lizard (p.153), which is larger, with a serrated collar.*

# Mosor Rock Lizard

*Lacerta mosorensis* (Squamata)

An elegant, long-tailed lizard, the Mosor Rock Lizard is a montane species, with a strongly flattened body that enables it to squeeze itself into small crevices. Within its restricted Balkan range, it tends to occupy more humid sites than other members of this family. Its habit of internally brooding its eggs for a period after fertilization may have evolved in response to the unreliable external temperature of its chosen habitat.

**FOUND** *in humid, shady, rocky mountain habitats, often foraging in open woodland and scrub.*

yellow, unspotted underparts

uniform mid-brown above

scattered irregular black speckles

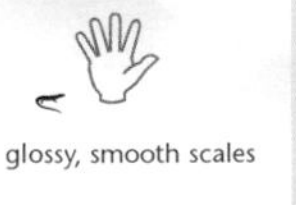

glossy, smooth scales

**DARKER FORM**

**SIZE** *Body up to 7cm; tail up to 15cm.*
**YOUNG** *Lays a single clutch of 4–8.*
**DIET** *Insects and other invertebrates.*
**STATUS** *Scarce.*
**SIMILAR SPECIES** *Hovarth's Wall Lizard (p.150); Sharp-snouted Rock Lizard (p.152), which has bluish underparts.*

# Sharp-snouted Rock Lizard

*Lacerta oxycephala* (Squamata)

**OCCUPIES** *a wide altitudinal range, from sea-level to mountains, usually in rather bare rocky areas, but also in towns.*

Occurring in two main colour forms, the Sharp-snouted Rock Lizard tends to be brown, dappled with paler spots, when its habitat is in lowland, but almost black when found in mountains. In both forms, its tail and underparts are blue, especially vivid in the male. A crevice-dweller with a flattened body, it is an agile climber, able to scale cliffs, rocks, and walls with ease.

often a blue tail, with dark rings

long head and pointed snout

blackish colour in montane form

DARK FORM

**SIZE** *Body up to 6.5cm; tail up to 10cm.*
**YOUNG** *Lays 2–4 eggs; hatch in 1–2 months.*
**DIET** *Insects and other invertebrates.*
**STATUS** *Locally common.*
**SIMILAR SPECIES** *Mosor Rock Lizard (p.151), which has yellowish underparts.*

# Meadow Lizard

*Darevskia praticola* (Squamata)

**FAVOURS** *moist and shady woodland, upland meadows and riverbanks; often basks on trees and stones.*

Similar in appearance to the Viviparous Lizard, the Meadow Lizard generally has a distinctive greenish or yellow belly. Also typical are its dark brown flanks contrasting with its pale brown back. Sometimes the edges of the colour bands are picked out by pale stripes or lines of dots. The Meadow Lizard can also be identified by its coarse scales, serrated collar, and white throat.

greenish or yellow belly

UNDERSIDE

**SIZE** *Body up to 6.5cm; tail up to 11cm.*
**YOUNG** *Lays 4–6 eggs; hatch after 1–2 months.*
**DIET** *Insects and other invertebrates.*
**STATUS** *Scarce.*
**SIMILAR SPECIES** *Viviparous Lizard (p.142); Horvath's Rock Lizard (p.150).*

# Balkan Wall Lizard

*Podarcis tauricus* (Squamata)

The Balkan Wall Lizard usually has a bright green back and red-brown flanks. It is, however, a very variable species, and may or may not have yellow side or flank stripes and black reticulation, or spotting on the flanks, sides of the belly, or back. It sometimes has a broken, black central stripe down the back. If seen clearly, it can be identified by its serrated collar.

**INHABITS** *lowland grassy habitats and field margins, scrub and vineyards, up to montane levels; hibernates in burrows or crevices.*

bright green back

**SIZE** *Body up to 8cm; tail up to 14cm.*
**YOUNG** *Lays up to six eggs.*
**DIET** *Insects and other invertebrates.*
**STATUS** *Common.*
**SIMILAR SPECIES** *Dalmatian Wall Lizard (p.151); Peloponnese Wall Lizard (p.154); Skyros Wall Lizard* (P. gaigeae).

COLOUR FORM

# Greek Rock Lizard

*Lacerta graeca* (Squamata)

With a flattened body and long head and tail, the Greek Rock Lizard is perhaps most distinctive in its lack of stripes. Its glossy grey-brown colour sometimes takes on bluish tones, especially in the male. It is extensively dappled with black spots on the back and pale brown and blue spots on the flanks. The pattern of the scales on its head is also characteristic. It has yellow-orange underparts, often with small black spots.

**LIVES** *in humid or damp areas in mountains, often in light shade, climbing rocks, cliffs, and trees with agility.*

pale spots on flanks

dark dappling on back

blue spots on flanks

grey-brown ground colour

♂

BREEDING MALE

plain tail

**SIZE** *Body up to 8cm; tail up to 18cm.*
**YOUNG** *Lays up to six eggs; hatch after 1–2 months.*
**DIET** *Insects and other invertebrates.*
**STATUS** *Locally common.*
**SIMILAR SPECIES** *Erhard's and Peloponnese Wall (p.154) Lizards, which are usually stripier.*

# Erhard's Wall Lizard

*Podarcis erhardii* (Squamata)

**AVOIDS** *human settlements, preferring dry lowland, rocky, and scrubby areas, and vegetated sand dunes.*

Although typically grey-brown and stripy, Erhard's Wall Lizard is very variable in the detail of its markings, particularly between different island populations. Many have been described as subspecies, but some may in fact merit recognition as species in their own right. When present, the dark side stripes are always stronger than any central stripe, a useful point of separation from the Common Wall Lizard, which occurs in parts of its range and shares broadly similar colours.

COLOUR VARIATION

**SIZE** *Body up to 7cm; tail up to 14cm.*
**YOUNG** *Up to four eggs.*
**DIET** *Insects and other invertebrates.*
**STATUS** *Locally common.*
**SIMILAR SPECIES** *Peloponnese Wall (below) and Greek Rock Lizards (p.153); Skyros* (P. gaigeae) *and Milos Wall Lizards* (P. milensis).

# Peloponnese Wall Lizard

*Podarcis peloponnesiacus* (Squamata)

**RANGES** *over the full array of Mediterranean habitats, from sea level to high in the mountains.*

One of the most robust of wall lizards, the male Peloponnese Wall Lizard is particularly well-built, with a noticeably large head. However, it is clumsy and lacks agility, so that it is restricted mainly to the ground or to walls that are not too precipitous. Although variable, it is usually strongly striped. Breeding males are often bright orange below, and have obvious blue spotting on the flanks, while juveniles sometimes have blue tails.

large head

strong yellow or green side and flank stripes

extensive blue spotting

♂

**SIZE** *Body up to 8.5cm; tail up to 15cm.*
**YOUNG** *Two clutches of up to six eggs.*
**DIET** *Insects and other invertebrates.*
**STATUS** *Common.*
**SIMILAR SPECIES** *Erhard's Wall Lizard (above); striped juvenile Balkan Green Lizard (p.139); Balkan Wall Lizard (p.153).*

# Tenerife Lizard

*Gallotia galloti* (Squamata)

Restricted to Tenerife and La Palma in the Canary Islands, the Tenerife Lizard is medium-sized, though often rather bulky. Males are generally very dark brown, in some forms with transverse bands of yellowish spots across the back, and usually with a blue throat and flank spots. In some cases, the blue spreads onto the chest and belly. The tail is reddish brown, similar to the overall ground colour of the female.

dark brown ground colour

♂

yellow eye

blue throat

blue flank spots

reddish brown ground colour

♀

pale flank stripe

**FOUND** *in all wild and cultivated habitats, apart from dense woodland and high mountain tops; frequently takes refuge in stone walls.*

**SIZE** *Body up to 15cm; tail up to 18cm.*
**YOUNG** *Up to two clusters of nine eggs.*
**DIET** *Fruit, other plant material, and insects.*
**STATUS** *Locally very common.*
**SIMILAR SPECIES** *Atlantic Lizard (below) and Boettger's Lizards* (G. caesaris), *on the islands of El Hierro and La Gomera.*

# Atlantic Lizard

*Gallotia atlantica* (Squamata)

A small, coarsely-scaled lizard, the Atlantic Lizard is the only lizard to be found on the islands it occupies naturally – Lanzarote and Fuerteventura, along with nearby minor islands. It is also found on Gran Canaria, but only as an introduction. Despite its relatively sombre colouration, it is very variable between localities: sometimes the throat is not blackish, and there are forms with different degrees of blue spotting on the flanks.

**INHABITS** *cultivated land, sand dunes, and old, vegetated lava fields, as well as areas around human habitation, in walls and gardens.*

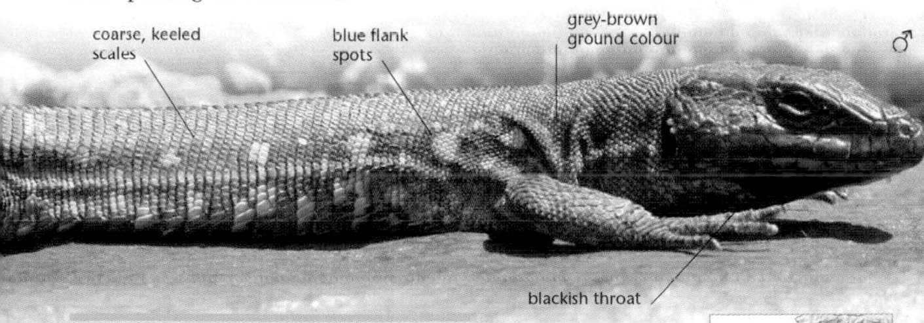


**SIZE** *Body up to 11cm; tail up to 15cm.*
**YOUNG** *Two or three clutches of up to five eggs.*
**DIET** *Insects, fruit, and flowers.*
**STATUS** *Locally common.*
**SIMILAR SPECIES** *Tenerife Lizard (above); Boettger's Lizard* (G. caesaris).

# Gran Canaria Giant Lizard

*Gallotia stehlini* (Squamata)

FOUND *in rocky, humid gorges, open woodland and scrub, and on mountains.*

Like the other large Canarian lizards, the maximum size of the Gran Canaria Giant Lizard has reduced by about a third since its island was colonized by humans. However, it remains a very large lizard. The scales on its back are strongly keeled and males have distinctive red throats, old males developing an almost black colouration. Adults have adopted a mainly vegetarian diet.

**SIZE** *Body up to 27cm; tail up to 50cm.*
**YOUNG** *Lays 4–16 eggs.*
**DIET** *Fruit, flowers, and leaves; also insects.*
**STATUS** *Locally common.*
**SIMILAR SPECIES** G. simonyi, G. intermedia, *and* G. bravoana, *on the islands of El Hierro, Tenerife, and La Gomera respectively.*

# West Canary Skink

*Chalcides viridianus* (Squamata)

OCCURS *in lowland areas, including cultivated land (even banana plantations), scrubby and stony hillsides, and pastures.*

Found on the Canary Islands of Tenerife, La Gomera, and El Hierro, the West Canary Skink has short limbs, which help it to climb. If it needs to move swiftly, it resorts to a slithering, snake-like motion. Its head is barely demarcated from its neck, and its scales are smooth and shiny. Its flanks are darker than its grey-brown ground colour.

**SIZE** *Body up to 9cm; tail up to 9cm.*
**YOUNG** *Up to six live young, in late summer.*
**DIET** *Insects and other invertebrates.*
**STATUS** *Very common.*
**SIMILAR SPECIES** *East Canary Skink* (C. simonyi), *is more uniform; Gran Canaria Skink* (C. sexlineatus), *is stripier.*

# Snake-eyed Skink

*Ablepharus kitaibellii* (Squamata)

Snake-like in appearance, with a transparent membrane covering the eyes instead of an eyelid, the Snake-eyed Skink has a cylindrical body, and shiny, bronze brown skin with darker flanks. When threatened, it withdraws its short limbs and wriggles to safety rather like a snake. It hides under logs and stones by day, emerging to hunt at dusk.

**PREFERS** *warm, dry, south-facing grassy slopes, fields, and woodland edges, with stone walls to bask on.*

**SIZE** *Body up to 5.5cm; tail up to 8cm.*
**YOUNG** *Lays 2–4 eggs in a hole in June; hatch in August–September.*
**DIET** *Spiders, flies, worms, and other invertebrates.*
**STATUS** *Common.*
**SIMILAR SPECIES** *None.*

# Bedriaga's Skink

*Chalcides bedriagai* (Squamata)

A glossy, bronze-tinted, short-legged lizard, Bedriaga's Skink usually has a series of small black spots with white centres (ocelli) running in bands across its back and broad tail. It has a transparent panel in the lower eyelid, enabling it to see even when its eyelids are closed, for example, while it is burrowing in soft sand.

**FOUND** *in dry, sandy, rocky habitats, often sparsely vegetated; also rough grassland and areas with dense leaf litter.*

indistinct head

elongated, cylindrical body

small, five-toed limbs

broad, tapering tail

**SIZE** *Body up to 8cm; tail up to 9cm.*
**YOUNG** *Two to three live young.*
**DIET** *Insects and other invertebrates.*
**STATUS** *Locally common, but secretive.*
**SIMILAR SPECIES** *Ocellated Skink (p.158), is more distinctly marked;* C. pistaceae, *which is a long-legged mountain form.*

# Ocellated Skink

*Chalcides ocellatus* (Squamata)

**OCCUPIES** *sandy, lowland habitats, especially near the sea, including scrubland, gardens, vineyards, and olive groves; takes refuge in holes in walls, under stones and logs, and in soft soil.*

A large skink, with a glossy, pale brown body, the most distinctive feature of the Ocellated Skink is its ocelli. These are black-edged white spots, which form bands across the back and tail. The western form, from Italy and Sardinia, is particularly boldly marked, with a pair of whitish stripes running either side of the mid-line, and heavily mottled flanks. The head is not distinct from the body, and likewise, the tapering tail is often as thick as the body for part of its length, lending a very characteristic cylindrical form to the species.

flanks heavily marked with black

cylindrical body

short limbs with five toes

WESTERN FORM

**NOTE**

*All skinks have reduced legs. Those of the Ocellated Skink are relatively large, and are used to clamber over rocks and stone walls; it uses serpentine wriggling to move quickly and to burrow in soft soil.*

smooth, glossy scales

pale brown ground colour

ocelli forming transverse bands

**SIZE** *Body up to 15cm; tail up to 15cm, but often shorter.*
**YOUNG** *Up to 20 live young; born in summer.*
**DIET** *Insects and other invertebrates.*
**STATUS** *Locally common.*
**SIMILAR SPECIES** *Bedriaga's Skink (p.157), which is smaller and less prominently marked; Levant Skink* (Euprepis auratus), *which has black flank blotches, and is found on some Aegean islands.*

# Western Three-toed Skink

*Chalcides striatus* (Squamata)

A very elongated lizard, the Western Three-toed Skink could be mistaken for a snake or slow-worm if the tiny, three-toed legs were not visible. Its cylindrical body, bearing 9–13 fine stripes, is distinctive. Hibernation takes place in holes in the ground, from which it emerges late in spring, as it cannot tolerate cold.

**INHABITS** *a range of habitats from damp but sunny meadows to dry coastal heaths and sand dunes.*

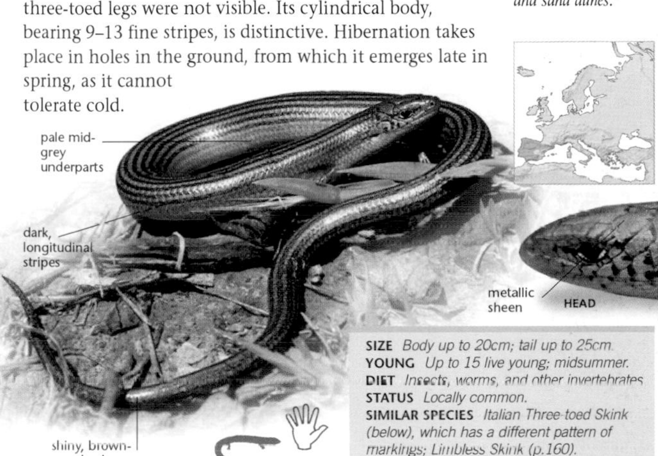


**SIZE** *Body up to 20cm; tail up to 25cm.*
**YOUNG** *Up to 15 live young; midsummer.*
**DIET** *Insects, worms, and other invertebrates.*
**STATUS** *Locally common.*
**SIMILAR SPECIES** *Italian Three-toed Skink (below), which has a different pattern of markings; Limbless Skink (p.160).*

# Italian Three-toed Skink

*Chalcides chalcides* (Squamata)

In its general appearance, the Italian Three-toed Skink, with its long, snake-like, metallic body, and tiny legs, is very similar to its relative, the Western Three-toed Skink. However, it differs in pattern with no markings at all or fewer relatively wide stripes. It has a pointed head but blunt snout, and its long middle hind toe is diagnostic.

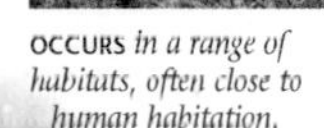

**OCCURS** *in a range of habitats, often close to human habitation.*

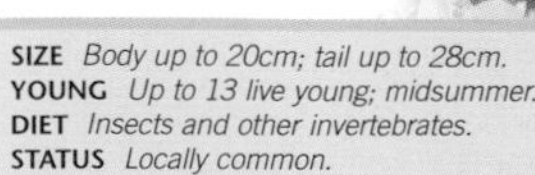

**SIZE** *Body up to 20cm; tail up to 28cm.*
**YOUNG** *Up to 13 live young; midsummer.*
**DIET** *Insects and other invertebrates.*
**STATUS** *Locally common.*
**SIMILAR SPECIES** *Western Three-toed Skink (above) has only 9–13 dark lines; Limbless Skink (p.160), Slow-worms (pp.161–162).*

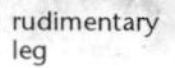

# Limbless Skink

*Ophiomorus punctatissimus* (Squamata)

**INHABITS** *dry, grassy slopes, with scattered stones under which it hides; burrows in soft earth, going deeper in dry weather.*

Resembling a small snake, the Limbless Skink has eyelids (snakes have none), showing that it is actually a lizard. It has small belly scales and a rather stiff body. Largely grey-brown, it may have a patterned tail. Since the tail is also detachable, the skink uses it in a defensive ploy. If threatened, it wriggles its tail furiously, drawing the predator's attention to the expendable part of its body.

**SIZE** *Up to 18cm.*
**YOUNG** *Poorly known; lays eggs.*
**DIET** *Insects and other invertebrates.*
**STATUS** *Very locally common.*
**SIMILAR SPECIES** *Western and Italian Three-toed Skinks (p.159); Slow-worms (pp.161–162); Worm-snake (p.163).*

# European Glass Lizard

*Pseudopus apodus* (Squamata)

**FAVOURS** *dry, grassy slopes, open woodland, and cultivated areas (including olive groves and vineyards).*

Large, snake-like, and legless (except for vestigial hind legs), the European Glass Lizard can grow to a considerable size, some specimens measuring up to 7cm in diameter. The male, in particular, has powerful jaws, used to crush its prey. Despite its bulk, it moves quickly, and both swims and climbs well. If shed, the regrown tail is often stumpy.

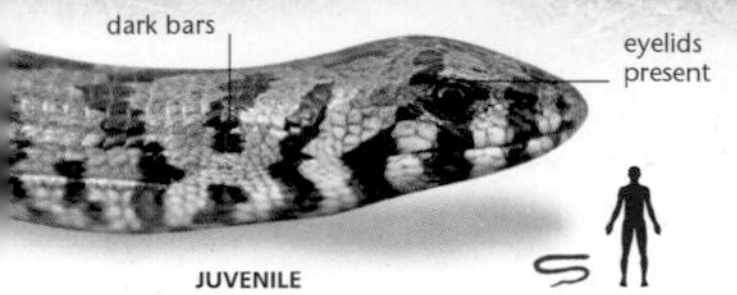


**SIZE** *1.2–1.4m.*
**YOUNG** *Six to ten eggs, laid June–July.*
**DIET** *Large insects, invertebrates, small reptiles, birds, and rodents.*
**STATUS** *Locally common.*
**SIMILAR SPECIES** *None; adults much larger than Slow-worms (pp.161–162).*

# Slow-worm

*Anguis fragilis* (Squamata)

Sometimes (incorrectly) called the Blind Worm, the Slow-worm is the most widely distributed legless lizard in Europe, where it is found throughout, except in the extreme north, south, and west. It has a blunt head and its underparts are paler than the grey-brown upperparts. The females sometimes have a dark stripe down the back, retained from the juveniles' colour, which is a deep bronze sheen. Unlike many reptiles, it rarely basks in the sun, apart from pregnant females encouraging the development of their young, and animals enticed to emerge from hibernation by warm winter weather.

**FOUND** *mostly in rather moist grassy areas, with scrub or hedgerows for refuge; also frequents large gardens and open woodland, up to montane levels.*

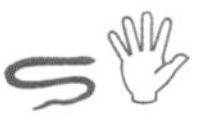

males sometimes have blue spots

uniform bronze grey-brown upperparts

snake-like, limbless body

♂

smooth scales

**NOTE**

*An important difference between legless lizards and snakes is that snakes can unhinge their jaws to tackle large prey; the Slow-worm cannot do this, restricting its diet to smaller prey.*

**SIZE** *30–50cm.*
**YOUNG** *Six to 12 (sometimes more) live young, born August–September; sometimes only every other year.*
**DIET** *Earthworms, insect larvae, spiders, slugs, and snails.*
**STATUS** *Common.*
**SIMILAR SPECIES** *Limbless Skink (p.160), which is smaller; Peloponnese Slow-worm (p.162), which is often longer.*

# Peloponnese Slow-worm

*Anguis cephallonica* (Squamata)

**FOUND** *in humid places in open woodland, along riverbanks, and up to moderately high altitudes.*

Very similar to the Slow-worm, the Peloponnese Slow-worm is usually longer, more slender, and less uniform in colour. It has sharply demarcated flanks and underparts, and sometimes, a short, dark stripe at the back of the neck.

shiny, mid-brown upperparts

dark brown flanks and underparts

wavy line behind head

**SIZE** *Up to 50cm.*
**YOUNG** *Live young, but little known.*
**DIET** *Earthworms, insect larvae, spiders, slugs, and snails.*
**STATUS** *Locally common.*
**SIMILAR SPECIES** *Limbless Skink (p.160), which is smaller; Slow-worm (p.161).*

narrow head

HEAD

# Iberian Worm Lizard

*Blanus cinereus* (Amphisbaenian)

**OCCURS** *in subterranean habitats, beneath cultivated land or pine woods.*

Also known as the Amphisbaenian, the Iberian Worm Lizard looks like a large, plump earthworm, sharing the worm's subterranean habits. Living and feeding underground, its vision is poor, but other senses are acute. It has yellow-brown to grey skin, often with pinkish tones. It can shed part of its blunt tail in an emergency, like a lizard, but it will not regrow.

worm-like segments

large protective scales for burrowing

HEAD

tiny eyes under skin

**SIZE** *Up to 30cm.*
**YOUNG** *Lays eggs.*
**DIET** *Insects, especially ants.*
**STATUS** *Possibly scarce, but poorly known.*
**SIMILAR SPECIES** *Anatolian Worm-lizard (B. strauchi), which is found only in the E. Mediterranean area.*

# Worm-snake

*Typhlops vermicularis* (Squamata)

A resident of the Balkans, the Worm-snake is a secretive small snake, living largely in underground burrows. Unlike all other European snakes, it lacks enlarged scales on its belly that are generally used to provide grip and so assist movement. In the Worm-snake, the spine on the end of its bulbous tail is believed to fulfil the same role. As with many subterranean animals, it has little need for vision, and the eyes have become reduced to minute organs, probably capable of detecting only the difference between dark and light.

**LIVES** *a subterranean life, in burrows under dry scrub or grazed fields; usually found by turning stones or occasionally on the surface at dusk or in wet weather.*

**NOTE**

*The swollen tail of the Worm-snake is believed to be a "false head" to attract the attention of its predators. Although this small snake is unable to shed its tail, its vital organs are concentrated towards its actual head.*

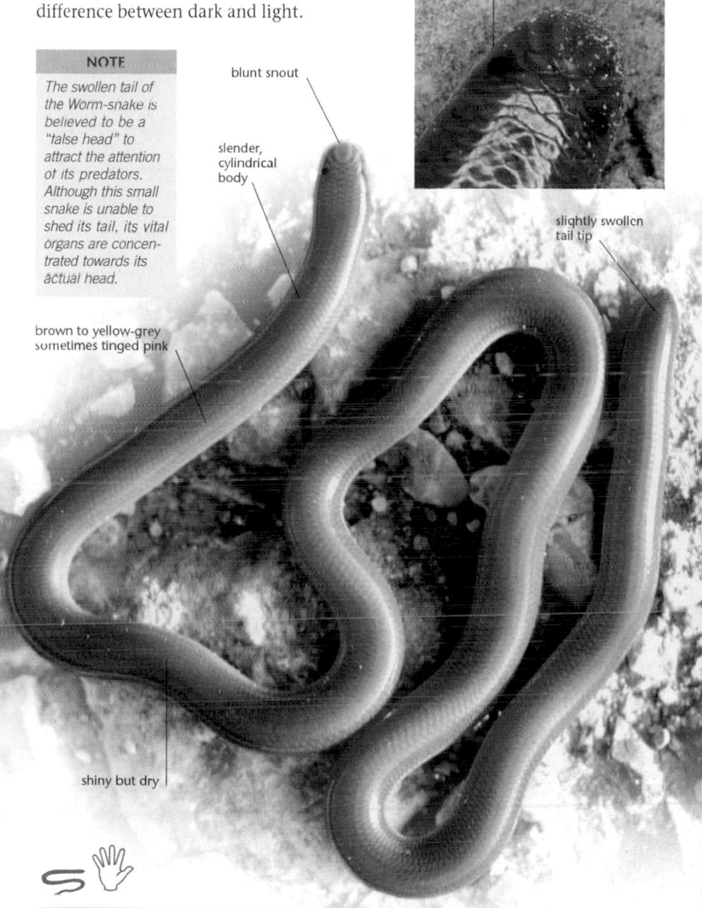


**SIZE** *Up to 40cm.*
**YOUNG** *Lays 4–8 eggs; poorly known.*
**DIET** *Small insects, especially ants, and other invertebrates.*
**STATUS** *Possibly scarce, but overlooked.*
**SIMILAR SPECIES** *Limbless Skink (p.160), which is more often found at the surface, although it burrows readily; Anatolian Worm-lizard* (B. strauchi), *which is more worm-like, with apparent segmentation.*

# Sand Boa

*Eryx jaculus* (Squamata)

**OCCURS** *on open or scrubby dry slopes, sand dunes and beaches, ploughed arable fields, and in dry river valleys.*

The only European member of the family with the largest snakes in the world, the Sand Boa is restricted to the Balkans. It has a characteristic body form – rather stocky, with a short, blunt tail and a pointed snout. Its yellow-brown ground colour, with dark brown blotches and bars, as well as blackish flank spots and head markings, provide excellent camouflage as it lies in wait for its prey. Largely nocturnal and crepuscular, the Sand Boa spends much of the day hidden under rocks or in rodent burrows.

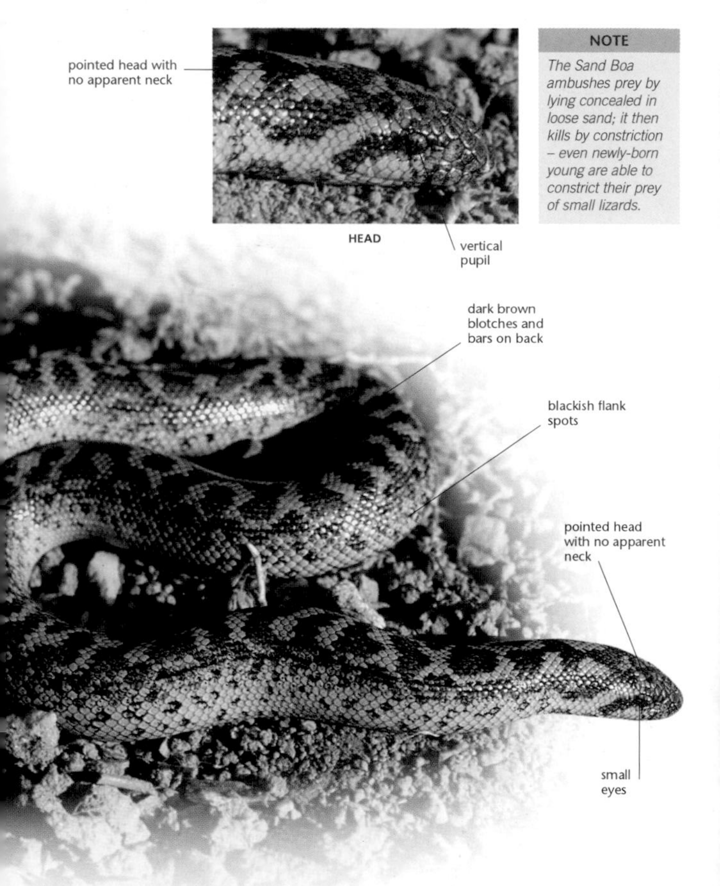


**NOTE**

*The Sand Boa ambushes prey by lying concealed in loose sand; it then kills by constriction – even newly-born young are able to constrict their prey of small lizards.*

**SIZE** *60–80cm.*
**YOUNG** *Up to 20 live young, August–September.*
**DIET** *Mice, voles, lizards, birds, and large invertebrates.*
**STATUS** *Scarce.*
**SIMILAR SPECIES** *Cat Snake (p.177).*

**FOUND** *in dry stony areas, from heathland to semi-desert, usually with scrub or stone walls for shelter; also near water in parts of its range.*

# Montpellier Snake

*Malpolon monspessulanus* (Squamata)

One of the longest European snakes, the Montpellier Snake is very slender, though rather stiff-bodied. The narrow head is barely thicker than the body, which is generally rather uniform in colour, ranging from sandy brown to almost black. Some adults, and particularly juveniles, show dark brown spotting, forming 5–7 rows down the back and flanks. It locates its prey by sight, lifting its head up and waving side-to-side "cobra-like". Once the prey is captured securely, the fangs inject the venom which kills in minutes. Humans, however, are unlikely to be bitten, and a bite is likely to cause only mild discomfort.

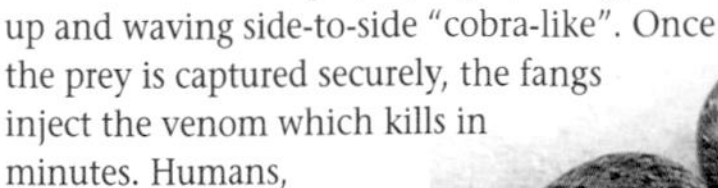

JUVENILE

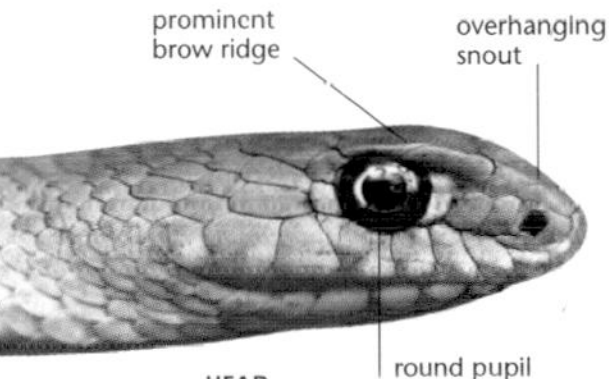


HEAD

**NOTE**

*Identification is easy once you see this snake's head: the raised "eyebrow" ridges, running forward onto the snout, combined with its large eyes, give it a "frowning" expression quite unique among European snakes.*

**SIZE** *Up to 2.4m.*
**YOUNG** *Up to 20 eggs, laid in holes in the ground in April–July, hatching September–October.*
**DIET** *Lizards, rodents, small Rabbits, other snakes, and large invertebrates.*
**STATUS** *Common.*
**SIMILAR SPECIES** *None.*

# Horseshoe Whip-snake

*Hemorrhois hippocrepis* (Squamata)

A slender, boldly marked snake, the Horseshoe Whip-snake is the only European member of its family to have a row of small scales immediately below the eye. Its ground colour is yellowish or olive, with a series of blackish purple diamonds running down the back. The head is marked with two or more dark bars, the rearmost often bent into the horseshoe shape from which it gets its name.

**FAVOURS** *dry, scrubby areas, usually with rocks, stone walls, or ruins, and around vineyards and olive and almond groves.*

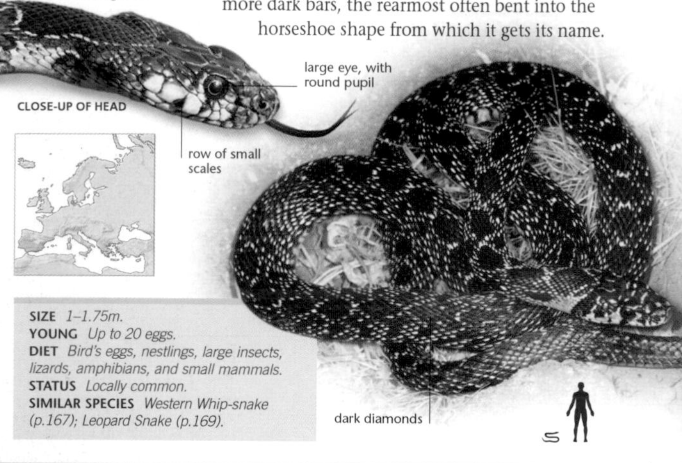


**SIZE** *1–1.75m.*
**YOUNG** *Up to 20 eggs.*
**DIET** *Bird's eggs, nestlings, large insects, lizards, amphibians, and small mammals.*
**STATUS** *Locally common.*
**SIMILAR SPECIES** *Western Whip-snake (p.167); Leopard Snake (p.169).*

# Dahl's Whip-snake

*Platyceps najadum* (Squamata)

Pale olive brown in colour, the very slender Dahl's Whip-snake is unmistakable across most of its Balkan range. A series of white-edged black eye-spots down the side of the neck are largest just behind the eye, and diminish in size rapidly further back. It also has a series of pale scales in front of and behind the eye. While not strictly venomous, its saliva seems to be toxic to its prey.

**OCCURS** *in dry stony areas and open woodland, mainly in the lowlands, where there are stone walls or rocks for refuge.*

**SIZE** *Up to 1.35m.*
**YOUNG** *Up to 16 eggs.*
**DIET** *Lizards, invertebrates, small mammals.*
**STATUS** *Scarce.*
**SIMILAR SPECIES** *Reddish Whip-snake* (P. collaris)*, which has larger white-edged spots; Dwarf Snake* (Eirenis modestus).

# Western Whip-snake

*Hierophis viridiflavus* (Squamata)

This slender, whip-like snake has a well-defined head. It is an efficient climber, scaling cliffs, rocks, and trees with ease. The Western Whip-snake hibernates, sometimes communally, in crevices, burrows, or buildings between October and March, but will emerge to bask on warm winter days. The typical form is a distinctive yellow-green colour, more-or-less covered by very dark blackish green blotches or bars, becoming longitudinal stripes on the tail. Juveniles are similar, though less boldly marked, apart from a strongly patterned head.

**PREFERS** *dry scrubby habitats, heathland, woodland, and gardens, usually with rock outcrops, scree or stone walls, and ruins.*

JUVENILE

striped tail

dark-coloured markings

yellow-green below

BLACK FORM

**NOTE**

*A variable species, whip-snakes from the eastern part of the range are often black; this is a useful distinction from the Balkan Whip-snake (p.168) which occurs in the same area, but is spotted.*

**SIZE** *Up to 1.8m.*
**YOUNG** *Lays 8–15 eggs; June–July; hatch after 1–2 months.*
**DIET** *Small lizards, large insects, rodents, and birds' nestlings.*
**STATUS** *Locally very common.*
**SIMILAR SPECIES** *Horseshoe Whip-snake (p.166); Balkan Whip-snake (p.168); black forms are like some Aesculapian Snakes (p.171); Algerian Whip-snake* (Hemorrhois Algirus), *which is paler.*

# Large Whip-snake

*Dolichopis caspius* (Squamata)

**OCCURS** *in open, dry habitats, from rocky hillsides to semi-desert; often basks on roads.*

A large Balkan species, the Large Whip-snake is less well marked than others of its group. Its olive brown upperparts have scales with pale centres, which form into indistinct stripes running the length of the body. The underparts are yellow or orange, and unspotted. Juveniles are more heavily marked, with dark spots and bars across the back, but unlike juvenile Western and Balkan Whip-snakes, they do not have clear yellow bars on the head.

JUVENILE

**SIZE** *Up to 2m.*
**YOUNG** *Lays 5–18 eggs.*
**DIET** *Small mammals and lizards; some large insects, small snakes, and birds.*
**STATUS** *Locally common.*
**SIMILAR SPECIES** *Balkan Whip-snake (below), which has spots as an adult.*

# Balkan Whip-snake

*Hierophis gemonenesis* (Squamata)

**PREFERS** *lowland, dry, stony, and scrubby areas, with ruins and stone walls.*

A slender, smooth-scaled snake, with a relatively uniform back, the adult Balkan Whip-snake usually has an array of spots, interspersed with white flecks, on the upperparts. These tend to fade out towards the tail. The underparts are pale yellow, with some dark spots towards the throat. Juveniles are more boldly marked, but difficult to differentiate from the Western Whip-snake.

JUVENILE

**SIZE** *1–1.5m.*
**YOUNG** *Up to 10 eggs.*
**DIET** *Lizards, small mammals, and insects.*
**STATUS** *Common.*
**SIMILAR SPECIES** *Large Whip-snake (above); Western Whip-snake (p.167); adults resemble Coin-marked Snake* (Hemorrhois nummifer).

# Leopard Snake

*Zamenis situla* (Squamata)

Formerly called *Elaphe situla*, this elegantly marked species comes in a variety of forms. Typically, its pale yellow-grey ground colour is patterned with black-edged red spots and blotches; sometimes these are joined into a pair of parallel red stripes running the entire length of the body. A docile snake, it climbs well on walls and in bushes, and hibernates in rock crevices between September and May. If threatened, it may vibrate its tail amongst dead vegetation, giving an audible warning, rather like a rattlesnake.

**INHABITS** *warm, dry stony and sandy areas, including steppe grassland, often with sparse shrubs, rocks, and stone walls.*

**NOTE**

*Juvenile snakes in the former genus* Elaphe *often have similar "leopard spots", though these are generally darker brown. Unlike most others, however, the Leopard Snake, retains this pattern as an adult.*

**STRIPED FORM**

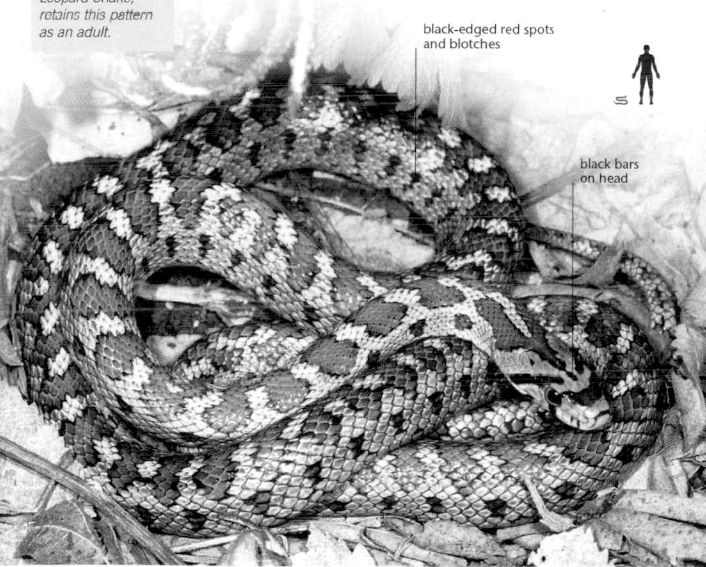


**SIZE** *80–110cm.*
**YOUNG** *Up to eight eggs, laid in a hole in the ground; hatch after 1–2 months.*
**DIET** *Rodents; also birds' eggs and nestlings, and lizards.*
**STATUS** *Scarce, perhaps declining.*
**SIMILAR SPECIES** *None as adults, within its range; Horseshoe Whip-snake (p.166), which has a separate, more western distribution.*

# Ladder Snake

*Rhinechis scalaris* (Squamata)

**FOUND** *on dry, south-facing slopes, with scrub and open woodland; also cultivated land.*

The Ladder Snake is a reddish olive colour, with two parallel dark lines running down its thick, powerful body. Juveniles are more distinctive, with H-shaped dark markings that are often fused to form the ladder for which it is named. An agile climber, it kills by constriction. It remains active by night in very warm weather. Hibernation takes place underground or in deep crevices, sometimes communally.

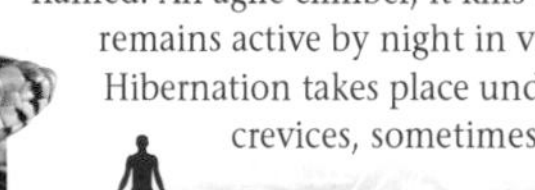


JUVENILE

**SIZE** *1.3–1.5m.*
**YOUNG** *Up to 15 eggs laid, June–July; hatch September–October.*
**DIET** *Rodents, lizards, birds, and birds' eggs.*
**STATUS** *Locally common.*
**SIMILAR SPECIES** *Four-lined Snake (below); Aesculapian snakes (p.171, 172).*

# Four-lined Snake

*Elaphe quatuorlineata* (Squamata)

**OCCUPIES** *dry, rocky hillsides, woodland edges, dry riverbeds, and steppes up to montane levels.*

Occuring in two distinct forms, the western Four-lined Snake has four dark lines as an adult, with keeled scales on its back, while the eastern form has bold dark blotches. Since they are so distinct, it is often suggested that they should be treated as two species, the eastern being the Blotched Snake *E. sauromates*. Both forms are large, robust, and powerful.

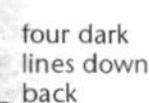

JUVENILE

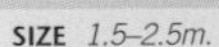

**SIZE** *1.5–2.5m.*
**YOUNG** *Lay 6–16 eggs; hatch in autumn.*
**DIET** *Rodents, birds' eggs and nestlings, and lizards, especially favoured by juveniles.*
**STATUS** *Locally common.*
**SIMILAR SPECIES** *Ladder Snake (above); Aesculapian snakes (p.171, 172).*

# Aesculapian Snake

*Zamenis longissimus* (Squamata)

The Aesculapian Snake is very slender and extremely agile, climbing trees, rocks, and cliffs with ease in search of prey and shelter. The overall appearance of the adult is generally uniform olive brown and shiny, due to the absence of keeled scales; it often has small pale flecks on the body. Sometimes, obscure dark stripes run down the back, and the underparts are whitish, often pale yellow on the throat. In contrast, juveniles are well marked with dark blotches and smooth keeled scales; they are similar to Grass Snakes with a pattern of yellow and black forming a partial collar on the back of the head.

**FAVOURS** *dry, open woodland and scrub, rocky slopes, stone walls and ruins; found up to montane levels, preferring south-facing slopes.*

**JUVENILE**

**DARK FORM**

**NOTE**

*Named after Aesculapius, the Greek god of healing, whose wooden staff was entwined with a snake, this species was encouraged around temples dedicated to him.*

relatively uniform olive brown upperparts

smooth, unkeeled scales

grey-brown eyes

**SIZE** *1–2m.*
**YOUNG** *Up to 18 eggs laid in holes in June–July; hatch in September.*
**DIET** *Rodents; some birds, amphibians, and other reptiles.*
**STATUS** *Common.*
**SIMILAR SPECIES** *Western Whip-snake (p.167); Four-lined Snake (p.170); Ladder Snake (p.170); Italian Aesculapian Snake (p.172). Grass Snake (p.174) similar to juveniles, which have smooth scales.*

# Italian Aesculapian Snake

*Zamenis lineatus* (Squamata)

**OCCUPIES** *dry, open woodland and scrub, rocky slopes, stone walls and ruins; avoids damp areas.*

Replacing the Aesculapian Snake in southern Italy and Sicily, the Italian Aesculapian Snake is typically paler, with pale grey underparts. Some adults show distinct dark brown lines running down the body. Other points of distinction from its more widespread relative include its red eye and relative lack of white flecks on the upperparts.

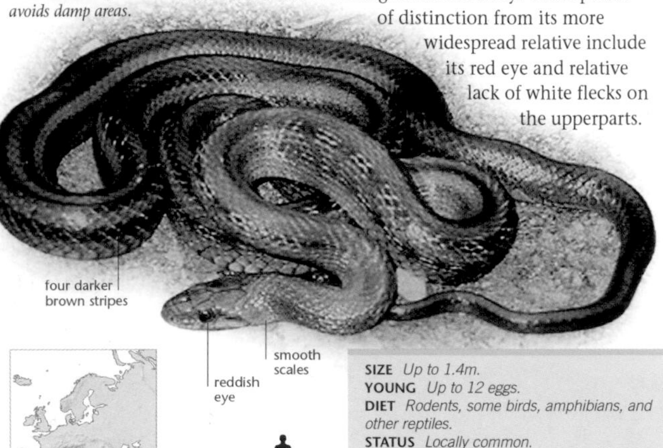


**SIZE** *Up to 1.4m.*
**YOUNG** *Up to 12 eggs.*
**DIET** *Rodents, some birds, amphibians, and other reptiles.*
**STATUS** *Locally common.*
**SIMILAR SPECIES** *Ladder and Four-lined Snakes (p.170); Aesculapian Snake (p.171).*

# Dice Snake

*Natrix tessellata* (Squamata)

**FAVOURS** *wetland habitats, in lowland, flowing and still waters; tolerant of brackish water by the coast.*

The most aquatic of European snakes, the Dice Snake is named after its pattern of dark, square blotches, each row offset to form a tessellated pattern. It often basks on stones or trees near water but, if disturbed, drops into the water and may remain submerged for several hours. Hibernation takes places in damp holes or crevices; these snakes emerge in April, males coming out before females.

pointed snout

grey ground colour

**SIZE** *75–120cm.*
**YOUNG** *Up to 25 eggs, July–August.*
**DIET** *Small fish, amphibians and aquatic invertebrates, ambushed in water.*
**STATUS** *Common.*
**SIMILAR SPECIES** *Viperine Snake (p.173); Grass Snake (p.174); Asp Viper (p.179).*

# Viperine Snake

*Natrix maura* (Squamata)

**FOUND** *close to waterbodies, from cold mountain streams to still, well-vegetated lowland ponds, marshes, and brackish coastal waters.*

The western counterpart of the Dice Snake, the Viperine Snake is a dark olive green colour, with dark brown spots, sometimes merged into a zig-zag stripe. The flanks often have ocelli, round blotches with a yellow centre. There may also be a dark V-shaped blotch on the back of the head, pointing forward, which adds to its viper-like pattern. The Viperine Snake has nostrils directed upwards. It hibernates only in the north of its range, in holes in the ground, hollow trees, or damp rock crevices.

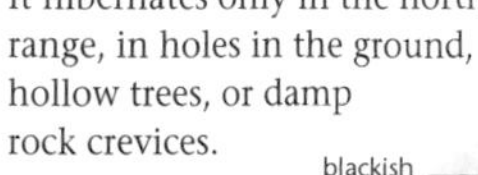

**NOTE**

*When threatened, a Viperine Snake hisses, adopting a threatening, viper-like posture, and strikes in defence. This is all a bluff, as it is not venomous and keeps its mouth closed when striking.*

HEAD

**SIZE** *70–100cm.*
**YOUNG** *Up to 20 eggs, May–June, hatching August–October.*
**DIET** *Frogs and other amphibians; some fish and invertebrates.*
**STATUS** *Common.*
**SIMILAR SPECIES** *Dice Snake (p.172); Grass Snake (p.174); animals with zig-zag marks resemble vipers (pp.178–181), which have vertical, not round, pupils.*

# Grass Snake

*Natrix natrix* (Reptilia)

**FAVOURS** *damp areas, feeding in rivers, ponds, and marshes; also in meadows, heathland, open woodland, and large gardens.*

Although not as aquatic as Dice and Viperine Snakes (pp.172–73), the Grass Snake swims well, with its head and neck out of the water (the nostrils are not set at the top of the snout), and does much of its hunting in water. The eggs are brooded in rotting vegetation, enabling the Grass Snake to be the most northerly egg-laying snake. Dark olive, with variable darker spotting, most Grass Snakes have a distinctive yellow collar, with a blackish rear border. However, in some southern subspecies, occasional melanic specimens, and old, faded individuals, the collar may not be present. Hibernation takes place between October and March, often in traditional communal sites.

STRIPED FORM

**NOTE**

*If captured, Grass Snakes first wriggle violently. If this fails to secure release, they squirt an evil-smelling fluid from their anal glands, and finally feign death, with mouth open and tongue hanging out.*

FEIGNING DEATH

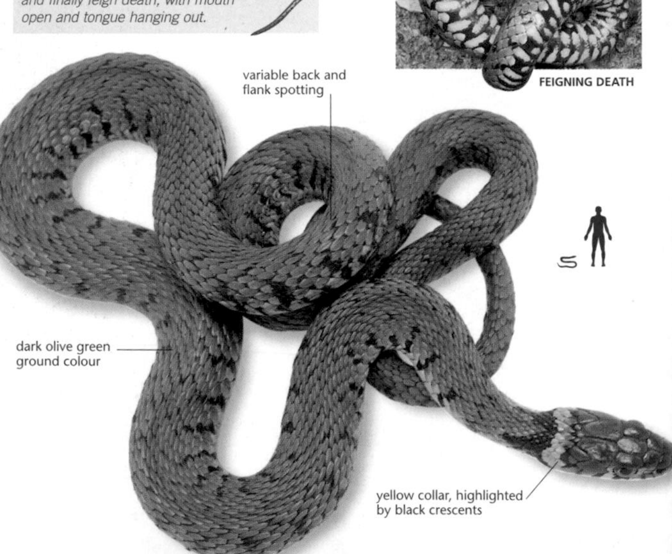

**SIZE** *70–150cm.*
**YOUNG** *Lays up to 100 eggs; hatch in August–September.*
**DIET** *Amphibians and fish pursued in water, small mammals, birds, lizards, and invertebrates.*
**STATUS** *Common.*
**SIMILAR SPECIES** *Dice (p.172) and Viperine (p.173) Snakes, which are more strongly aquatic and lack the yellow neck collar.*

# Smooth Snake

*Coronella austriaca* (Reptilia)

A secretive, slow-moving snake, the Smooth Snake spends much of its time underground in burrows. As a result, most surveys concentrate on locating sloughed skins. It is declining in parts of its range through habitat fragmentation. As its name suggests, it has a smooth appearance, due to the absence of raised keels on the scales on its back. Its generally uniform grey-brown colour is usually marked with paired small brown spots running down the back, a larger blotch (often heart-shaped) behind the head, and a black line running from the neck, through the eye, to the nostrils.

**FOUND** *in dry, sunny heathland habitats; also hedgerows and open woodland, up to montane levels in the south.*



**NOTE**

*Smooth Snakes are often easier to see than other snakes. When basking in the sun (they avoid the hottest part of the day), they tend to freeze, rather than flee, if disturbed.*

**SIZE** *50–70cm.*
**YOUNG** *5–16 live young; born August–September.*
**DIET** *Rodents and shrews; also lizards and some invertebrates.*
**STATUS** *Locally common; scarce and declining in the north.*
**SIMILAR SPECIES** *Southern Smooth Snake (p.176), which is blotched below, and the black facial stripe does not reach the nostrils; False Smooth Snake (p.176), which often has a black collar.*

# Southern Smooth Snake

*Coronella girondica* (Squamata)

**OCCUPIES** *warm, dry heathland, rocky slopes, Mediterranean scrub, and olive and almond groves; favours stone walls.*

Overlapping in distribution with the Smooth Snake, the Southern Smooth Snake is a largely nocturnal or crepuscular hunter that rests by day (and hibernates) in burrows and rock crevices. On the back of its head is a horseshoe-shaped blotch, and a dark stripe from its neck to the eye, not extending forward to the nostrils. Its upperparts are marked with darker spots and bars, which are fused into a viper-like zig-zag line, while its underparts are yellow or red.

grey-brown upperparts

dark spots or bars on back

slender body

large black blotches

UNDERSIDE

**SIZE** *Body up to 90cm.*
**YOUNG** *Lays up to 16 eggs; hatch after 2–3 months.*
**DIET** *Lizards and large invertebrates.*
**STATUS** *Locally common.*
**SIMILAR SPECIES** *False Smooth Snake (below); Smooth Snake (p.175).*

# False Smooth Snake

*Macroprotodon cucullatus* (Squamata)

**FAVOURS** *lowland, hot and dry habitats, such as open woods, cork oak plantations, and scrub, usually on rocks or stone walls.*

A largely nocturnal hunter, the False Smooth Snake is a pale grey-brown colour, with variable but small dark spots down the back. Usually, there is a black or dark brown collar, which is sometimes enlarged into a complete hood, and the facial stripe extends forward to the nostrils. Its underparts are white or yellow, sometimes with a pink tinge, and usually with dark blotches or stripes.

HEAD

**SIZE** *Body up to 65cm.*
**YOUNG** *Lays up to six eggs, under stones or in sandy ground; breeds every other year.*
**DIET** *Lizards, small mammals, and birds.*
**STATUS** *Locally common.*
**SIMILAR SPECIES** *Southern Smooth Snake (above); Smooth Snake (p.175).*

UNDERSIDE

# Cat Snake

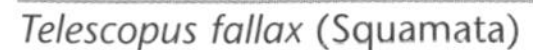

*Telescopus fallax* (Squamata)

Slender and laterally compressed, the Cat Snake is named for its cat-like vertical pupils (rounded in dim light) and its habit of stalking prey in open ground. Although venomous, its mouth is too small for its fangs to be used effectively on humans. It is active day and night, but is mainly a crepuscular hunter in the hottest part of the year. Black blotches are usually present, except on some island forms.

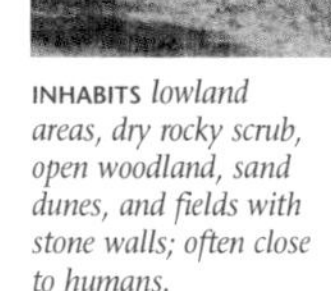

**INHABITS** *lowland areas, dry rocky scrub, open woodland, sand dunes, and fields with stone walls; often close to humans.*

**SIZE** *Body 60–100cm.*
**YOUNG** *Lays 5–9 eggs in June–July; hatch in September.*
**DIET** *Lizards, insects, and small mammals.*
**STATUS** *Locally common.*
**SIMILAR SPECIES** *Sand Boa (p.164), which is fatter; Vipers (pp.177–181).*

# Orsini's Viper

*Vipera ursinii* (Squamata)

Also called the Meadow Viper, Orsini's Viper is the smallest European member of its family. Its relatively slender body is variably coloured, but has a wavy, dark, black-edged stripe down the back. The stripe is dark brown in the male, and reddish in the female. A number of subspecies have different head shapes, colours, and patterns, and may be better treated as separate species. Orsini's Viper hibernates, and is venomous, but not aggressive.

**FOUND** *in lowland plains and steppes, often lacking scrub or tree cover, sometimes on the edge of marshes and on mountains.*

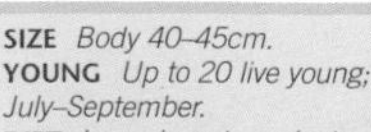

**SIZE** *Body 40–45cm.*
**YOUNG** *Up to 20 live young; July–September.*
**DIET** *Large insects, rodents, and lizards.*
**STATUS** *Endangered; locally common.*
**SIMILAR SPECIES** *Adder (p.178); Asp Viper (p.179); Nose-horned Viper (p.181).*

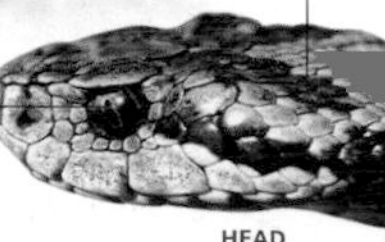

# Adder

## *Vipera berus* (Squamata)

**PREFERS** *heathland and moorland, open woods, meadows, and marshes; largely montane in the south.*

With a robust body (especially females), the Adder is a cold-tolerant species, found well into the Arctic Circle, and is the most northerly snake in the world. Its grey body, sometimes with yellow or reddish tones, is strongly marked with a dark stripe down the back, black in males and brown in females. Although it basks in sunshine, it tends to avoid the hottest conditions, and in the south hunts mostly at dusk. Some populations, especially in the Balkans, frequently produce melanic specimens – all black, with bright red eyes. Adders are quite venomous and bites require medical attention.

♀

**NOTE**

*Adders hibernate between September/ October and February/March (according to location). Traditional hibernation sites may be used communally, and adults can navigate to them from distances of 2km or more.*

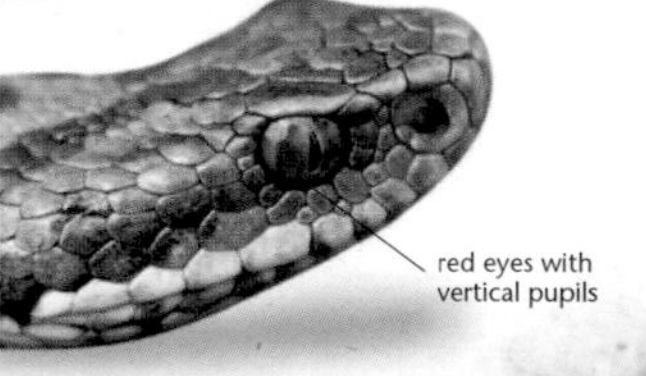


HEAD

dark grey to black underside

♂

dark V-mark on back of head

dark zig-zag stripe

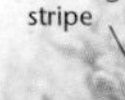

**SIZE** *50–65cm.*
**YOUNG** *Up to 20 live young in August, though not breeding every year.*
**DIET** *Small mammals (pursued in their burrows), lizards, newts, frogs, and nestling birds.*
**STATUS** *Common.*
**SIMILAR SPECIES** *Orsini's Viper (p.177); Seoane's Viper (p.179); Asp Viper (p.179); Nose-horned Viper (p.181).*

# Seoane's Viper

*Vipera seoanei* (Squamata)

Taking the place of the Adder in the north-west of the Iberian Peninsula, Seoane's Viper is similar to the more widespread species, and until recently was considered only a subspecies. The pattern of dark markings varies from a straight or wavy line (often flanked by pale stripes), to a series of cross bars, or even all black; the brown marks generally have a black margin. Its snout is sometimes slightly upturned.

**FAVOURS** *scrub, rough grassland, woodland edges, and clearings, from lowland to montane levels; avoids hot, dry, rocky sites.*

beige-grey ground colour

variable back stripe or bars

BLACK FORM

**SIZE** *Up to 65cm.*
**YOUNG** *Up to 10 live young.*
**DIET** *Small mammals, reptiles, amphibians, and birds.*
**STATUS** *Common.*
**SIMILAR SPECIES** *Asp Viper (below); Adder (p.178), which has a different range.*

# Asp Viper

*Vipera aspis* (Squamata)

The slightly upturned snout and a dark V or triangle on the back of the head are about the only consistent features of the Asp Viper. The back markings may be a series of blotches or transverse bars, sometimes linked by a black or dark brown stripe. Occasionally, an Adder-like zig-zag may be present, sometimes with a pale grey centre.

**OCCUPIES** *warm, dry areas in mountains, rocky places, and agricultural land in adjacent lowland.*

pale grey ground colour

BLACK FORM

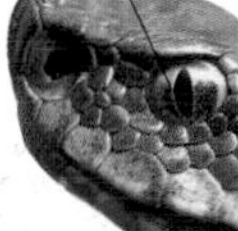

upturned snout

golden (or red) eyes

HEAD

**SIZE** *50–60cm.*
**YOUNG** *Up to 20 live young.*
**DIET** *Rodents, shrews, lizards, and birds.*
**STATUS** *Common.*
**SIMILAR SPECIES** *Seoane's Viper (above); Dice Snake (p.172); Orsini's Viper (p.177); Adder (p.178); Nose-horned Viper (p.181).*

# Lataste's Viper

*Vipera latasti* (Squamata)

**FOUND** *in dry, sunny habitats, including open woods, sand dunes, scrub, hedgerows, and stone walls; often in upland areas, to high mountain levels.*

The widespread viper of the Iberian Peninsula, Lataste's Viper normally shares the soft, elongated nose-horn of the Balkan Nose-horned Viper; in those individuals lacking the projection, the snout is distinctly upturned. Generally grey-brown above, sometimes with red or orange tones, the dark brown back markings consist of a zig-zag stripe, or a series of lobes, usually linked with a line. In either case, the markings are usually outlined in black. There is also a series of distinct spots or bars on the flanks.

grey-brown ground colour

dark brown markings

spots on flank

### NOTE

*The tip of the snout is of great value in identifying vipers. Nose-horned and Lataste's have a distinct projection; an upturned snout indicates Asp or (weakly upturned) Seoane's; while flat snouts are found on Orsini's Viper and the Adder.*

distinct nose projection

black facial bar

**HEAD PROFILE**

**SIZE** *Up to 60cm.*
**YOUNG** *Four to nine live young, not usually every year.*
**DIET** *Small mammals, lizards, snakes, birds, amphibians, and large invertebrates.*
**STATUS** *Locally common, but declining in the lowlands.*
**SIMILAR SPECIES** *Nose-horned Viper (p.181), which has an eastern distribution; hornless Lataste's are similar to Asp and Seoane's (p.179).*

# Nose-horned Viper

*Vipera ammodytes* (Squamata)

The largest and most dangerous European viper, the Nose-horned Viper is distinguished by the distinct, horn-like projection (up to 5mm long) from its snout. Although it frequently basks in the open, it avoids very hot conditions by moving to more shady areas, or adopting crepuscular or nocturnal hunting habits. Its stocky body is generally pale grey to yellowish brown, with a black-edged, dark brown wavy or zig-zag line, or a series of usually connected lobes down the back. Hibernation takes place in rock crevices, between October and March, although some emerge to bask in the midwinter sun.

red-brown stripe

♀

**FOUND** *on dry, rocky hillsides with scattered scrub, woodland margins, and clearings, vineyards and stone walls, from lowland to montane levels.*

**NOTE**

*Although this common venomous snake may be sluggish, it strikes quickly and effectively. However, this usually occurs only after persistent molestation. Prompt medical treatment is essential in case of a bite.*

horn-like projection

**HEAD PROFILE**

indistinct dark flank spots

pale grey to brown ground colour

variable black-edged, dark brown stripe

♂

**SIZE** *65–80cm.*
**YOUNG** *Up to 15 live young, August–September.*
**DIET** *Small mammals, birds, lizards, and small snakes.*
**STATUS** *Scarce, as a result of persecution.*
**SIMILAR SPECIES** *Asp Viper (p.179); Lataste's Viper (p.180); other eastern vipers are Orsini's Viper (p.177), Adder (p.178), Ottoman Viper* (Montivipera xanthina), *and Milos Viper* (Macrovipera schweizeri).

# Amphibians

Almost all European amphibians (like the Common Tree-frog pictured below) are dependent on fresh water, at least for breeding, and are usually found in moist habitats. The European total of about 75 amphibian species compares with a world total of about 5,000, a number that is rising as new species are discovered, especially in moist tropical areas. This book covers 53 European species. With the advent of novel techniques of genetic analysis, the relationship between many species and forms is being reconsidered. Thus, new species are continuously being recognized and the list continues to grow.

GOLDEN STRIPED SALMANDER

PAINTED FROG

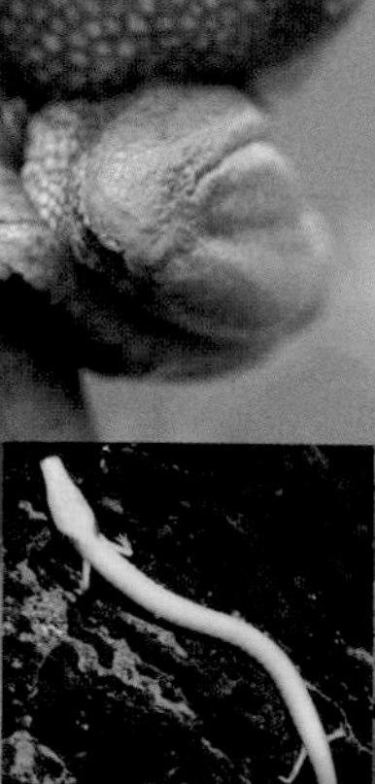

OLM

ITALIAN CRESTED-NEWT

# Fire Salamander

*Salamandra salamandra* (Caudata)

The bright colours of the Fire Salamander warn potential predators that it is poisonous: glands under the skin, and the raised parotid glands behind the eyes, secrete an irritant liquid. It is long-lived, sometimes up to 40 years. Although generally slow and lumbering, it lunges at prey with surprising speed. It hibernates under leaf mould or logs between October and March, when several salamanders may be found together, the only time apart from mating when they are not solitary.

**OCCUPIES** *damp, shady deciduous woodland, often in upland areas, hiding by day under logs; largely terrestrial but breeds in shady streams and pools.*

pattern of yellow or orange blotches

SPOTTED FORM

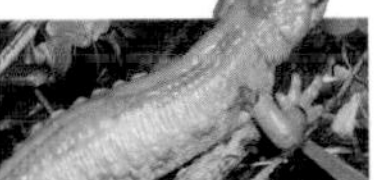

large parotid glands

STRIPED IBERIAN FORM

broad head

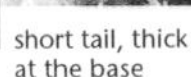

short tail, thick at the base

YELLOW FORM

short tail, thick at base

thick limbs

**NOTE**

*Fire Salamanders are variable in the extent and colour of blotching. Some forms, such as the Pyrenean subspecies, have blotches fused into two parallel lines; others, such as the Italian subspecies, are almost all yellow.*

**SIZE** *Body up to 14cm; tail up to 8cm.*
**YOUNG** *Up to 80 live young born in water in summer; metamorphose after 1–4 months.*
**DIET** *Worms, slugs, insects, and other invertebrates.*
**STATUS** *Common.*
**SIMILAR SPECIES** *Corsican Fire Salamander (p.184); Lanza's Alpine (p.184) and Alpine (p.185) Salamanders, which are primarily black.*

# Corsican Fire Salamander

*Salamandra corsica* (Caudata)

**PREFERS** *damp, shady woodland, gorges, and caves, often in upland areas, hiding by day under logs and rocks; breeds in shady streams and pools.*

The Corsican Fire Salamander is more aquatic than the similar Fire Salamander, with partially webbed toes and a laterally flattened tail, used for swimming. Generally with a yellow pattern, the blotches are often much reduced, in common with some forms of its more widespread relative. The parotid glands, while still prominent, are also usually smaller than those of the Fire Salamander.

**SIZE** *Body up to 14cm; tail up to 8cm.*
**YOUNG** *Up to 60 live young born in water.*
**DIET** *Worms, slugs, insects and other invertebrates, hunted by night.*
**STATUS** *Locally common.*
**SIMILAR SPECIES** *Fire Salamander (p.183), which does not occur on Corsica.*

# Lanza's Alpine Salamander

*Salamander lanzai* (Caudata)

**FOUND** *in montane areas of meadows, streams, screes, and woodland.*

A mountain-dwelling species, Lanza's Alpine Salamander is very similar to the all-black forms of Alpine Salamander, but with a more restricted distribution; the two species are not found together. Otherwise, Lanza's is a little larger, with a flatter head, and longer tail, although the latter has a distinctively rounded tip. Mainly nocturnal, it is active by day in wet weather.

flattened head
shiny black skin
prominent ribs
rounded tail tip
slight webbing between toes

**SIZE** *Body up to 9cm; tail up to 8cm.*
**YOUNG** *Up to six fully developed young.*
**DIET** *Worms, slugs, and other invertebrates.*
**STATUS** *Vulnerable; locally common.*
**SIMILAR SPECIES** *Fire Salamander (p.183); Alpine Salamander (p.185), which has a finely pointed tail-tip, but a shorter tail.*

# Alpine Salamander

*Salamandra atra* (Caudata)

Like a Fire Salamander without the "flames", the Alpine Salamander is usually black and glossy all over, although one Italian form has pale, often yellowish, patches extending down the back and tail, and on to the legs. Always secretive, generally nocturnal and hibernating for up to eight months of the year, Alpine Salamanders are likely to be seen only in wet weather, when they may be active during the day, or in their places of refuge – under logs or stones, in holes and crevices, or deep, damp leaf-litter, usually in mountainous areas.

**OCCURS** *in high mountain areas, as high as 2,800m, favouring damp, mossy woodland and stony, alpine meadows and heathland.*

**NOTE**

*Unlike most amphibians, this salamander does not need to breed in water. After developing for two or three years, the young are born fully formed (metamorphosed) as miniature versions of the adults, on land in damp areas.*

ribbed appearance

ITALIAN FORM

**SIZE** *Body up to 8cm; tail up to 7cm.*
**YOUNG** *Up to three live young, up to 5cm long.*
**DIET** *Worms, slugs, insects, and other invertebrates.*
**STATUS** *Locally common.*
**SIMILAR SPECIES** *Fire Salmander (p.183), which even in its darkest form, has some yellow/orange spotting; Lanza's Alpine Salamander (p.184), which has a longer but blunt tail, and a different range.*

# Golden-striped Salamander

*Chioglossa lusitanica* (Caudata)

**FOUND** *in damp mountainous areas, usually near rocky streams; shelters under stones and moss.*

A very secretive species, usually found in remote areas, the Golden-striped Salamander is active only at night and hibernates in winter. Dark brown, with grooves down its back and flanks, it is superficially lizard-like, with a slender tail more than twice the length of its body. Unusually for a salamander, it moves rapidly, and can shed and re-grow its tail.

two parallel coppery lines down back

long, cylindrical tail

smooth, glossy skin

**SIZE** *Body up to 5cm; tail up to 11cm.*
**YOUNG** *Lays 12–20 eggs, attached to rocks in shallow water; hatch in 2–3 months.*
**DIET** *Insects, worms, spiders, and other invertebrates.*
**STATUS** *Near-threatened; locally common.*
**SIMILAR SPECIES** *None.*

# Spectacled Salamander

*Salamandrina terdigitata* (Caudata)

**OCCUPIES** *shady mountain areas and north-facing hill slopes, often near rocky streams.*

A rather slender species, the Spectacled Salamander is the only European salamander to have four toes on its hind feet. It has undistinguished black, ribbed upperparts, broken only by a yellowish triangle on its head, but its underparts are an unmistakable combination of black, white, and vivid red. If threatened, it will loop its tail over its back, revealing the red beneath as a warning of its toxic skin secretions.

ribbed appearance

yellowish triangle on head

black upperparts

pale belly with black blotches

black throat

UNDERSIDE

red under tail and legs

**SIZE** *Body up to 5cm; tail up to 6cm.*
**YOUNG** *Lays 30–60 eggs, attached to rocks and debris in water.*
**DIET** *Invertebrates, including snails and slugs, in damp habitats.*
**STATUS** *Scarce and local.*
**SIMILAR SPECIES** *None.*

# Italian Cave-salamander

*Speleomantes italicus* (Caudata)

Cave-salamanders are a closely related group of six or seven species in Europe, each found in a distinct area: the Italian Cave-salamander is the species of the Italian peninsula. Variable in colour, with dark underparts, it may be difficult to identify. As with all cave-salamanders, it has no lungs and gets the oxygen it needs from diffusion through its moist skin.

partly webbed feet

prominent eyes

brown back, marbled red or yellow-brown

**FAVOURS** *humid, shady rocky areas and caves, usually on limestone, often around shaded streams running through mossy woodland.*

**SIZE** *Body up to 6.5cm; tail up to 6cm.*
**YOUNG** *Lays 5–10 eggs on land.*
**DIET** *Small invertebrates.*
**STATUS** *Near-threatened; common within its limited range.*
**SIMILAR SPECIES** *Other cave-salamanders (below and p.188).*

# Ambrosi's Cave-salamander

*Speleomantes ambrosii* (Caudata)

Although generally quite dark, with pale spots on its underparts, the variation within this species is such that colour alone cannot provide a definitive identification. Its range overlaps marginally with the Italian Cave-salamander and, in that area, hybrids occur, adding to identification difficulties. Like all cave-salamanders, the female guards her eggs from the cannibalistic habits of her species, although she is believed to eat some of her own young during the long incubation period of up to a year.

prominent eyes

partly webbed feet

dark brown back, variably marked with paler patches

**LIVES IN** *humid, shady rocky areas and caves, often near shaded streams running through mossy woodland.*

**SIZE** *Body up to 6.5cm; tail up to 6cm.*
**YOUNG** *Lays 5–10 eggs on land.*
**DIET** *Small invertebrates.*
**STATUS** *Near-threatened; locally common.*
**SIMILAR SPECIES** *Other cave-salamanders (above and p.188); western animals sometimes called* S. strinatii.

# Supramontane Cave-salamander

*Speleomantes supramontis* (Caudata)

**OCCURS** *in humid rocky areas, in mountains and their foothills, in caves, and near shaded, mossy streams in woodland.*

As a result of its range, restricted to a tiny part of eastern Sardinia, the Supramontane Cave-salamander is considered to be endangered globally. Even within its range, it is generally rather scarce, and may currently be declining. It has pale underparts, with dark spots. Given the variability in colour and pattern within each cave-salamander species, the main way of identifying them, even between the four Sardinian species, is distribution.

prominent eyes

variably marbled brown back

**SIZE** *Body up to 7cm; tail up to 6cm.*
**YOUNG** *Lays 5–10 eggs in damp crevices.*
**DIET** *Small invertebrates.*
**STATUS** *Endangered; generally scarce.*
**SIMILAR SPECIES** *Monte Albo Cave-salamander* (S. flavus), *which is yellower and larger, and found in NE Sardinia.*

# Scented Cave-salamander

*Speleomantes imperialis* (Caudata)

**INHABITS** *humid, shady rocky areas and caves, often around shaded streams in mossy woodland.*

Although it has widest range among the Sardinian cave-salamanders, the Scented Cave-salamander is still a very restricted species, found only in an area in the east of Sardinia. Apart from distribution, its most distinctive feature is the fact that when handled it produces a strong, aromatic smell. It has pale, usually unspotted, underparts.

**SIZE** *Body up to 6.5cm; tail up to 6cm.*
**YOUNG** *Lays 5–10 terrestrial eggs; some forms may bear live young.*
**DIET** *Small invertebrates.*
**STATUS** *Near-threatened; locally common.*
**SIMILAR SPECIES** *Gene's Cave-salamander* (S. genei), *which is smaller and darker.*

# Olm

*Proteus anguinus* (Caudata)

A truly bizarre creature, the Olm has an eel-like form, adapted to swimming in its subterranean water habitat. Its large tail also contributes to its swimming ability. As with many undergound animals, it has very little skin pigmentation and rudimentary vision, its skin-covered eyes incapable of little more than differentiating light from dark. It has three toes on the front foot and only two on the hind foot. But most remarkable are its three pairs of feathery external gills, which it retains throughout its life. In most amphibians, the gills are absorbed as the larva develops, but in the Olm, it is the larval form that assumes sexual maturity, in a phenomenon known as neoteny.

**FAVOURS** *subterranean waters, down to 300m underground, in caves and mines in limestone areas; occasionally coming to the surface in springs and in cave entrances.*



**NOTE**

*Living only in subterranean waters, the Olm is likely to be seen only in caves, although it can be flushed to the surface in storms; prolonged exposure to light turns its skin blackish.*

**SIZE** *Body up to 22cm; tail up to 8cm.*
**YOUNG** *Produces 1–2 live young; around 70 eggs may be laid.*
**DIET** *Small aquatic crustaceans.*
**STATUS** *Vulnerable; rare and secretive, discovered as recently as 1875.*
**SIMILAR SPECIES** *None.*

# Sharp-ribbed Newt

*Pleurodeles waltl* (Caudata)

**OCCUPIES** *lowland, still water habitats, including ponds, lakes, and ditches; woodland and cultivated land.*

The largest tailed amphibian in Europe, the Sharp-ribbed Newt is heavily built, with grey-brown, granular skin. Most distinctive is a row of orange wart-like poison glands on the flanks at the end of each rib, the tip of the rib protruding through the gland. Strongly aquatic, these newts rarely leave water. They are most active by night or during the day in rainy weather.

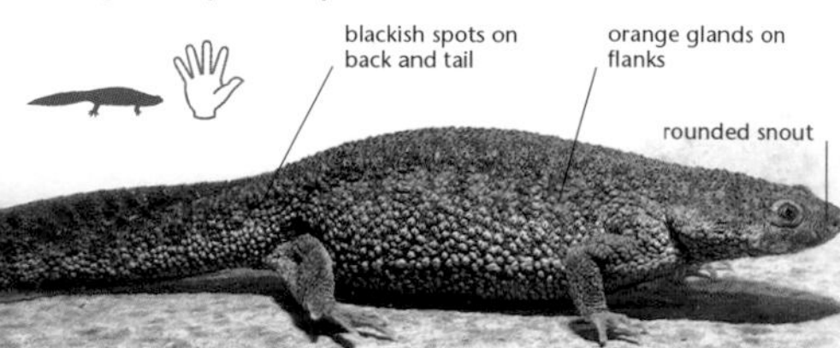


**SIZE** *Body up to 15cm; tail up to 15cm.*
**YOUNG** *Lays 200–1,000 eggs; throughout the year, in aquatic vegetation.*
**DIET** *Insects, worms, fish, frogs, and newts.*
**STATUS** *Locally common.*
**SIMILAR SPECIES** *Pyrenean Brook Newt (below), which does not have warty flanks.*

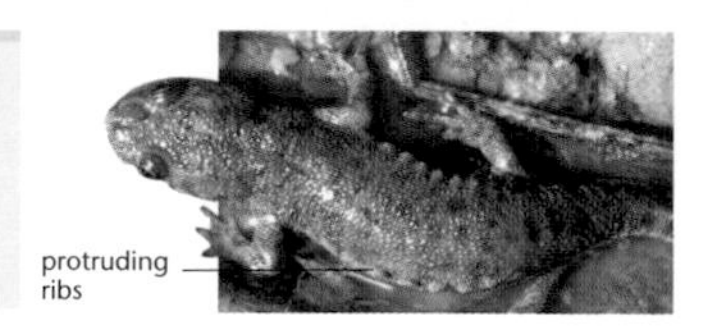


# Pyrenean Brook Newt

*Euproctus asper* (Caudata)

**FAVOURS** *cold mountain streams and lakes with stony beds; sometimes in caves.*

Restricted to the Pyrenees, and unlikely to be confused with any other species within its range, the Pyrenean Brook Newt has very rough skin, each granule with a hardened tip. It is active by day and night, but leaves the water only under the cover of darkness; it hibernates at high mountain levels, but aestivates at lower altitudes.

grey to olive brown above

rough granular skin

flank spots

STRIPED FORM

**SIZE** *Body up to 8cm; tail up to 7cm.*
**YOUNG** *Lays 20–40 eggs; in rock crevices.*
**DIET** *Invertebrates.*
**STATUS** *Locally common.*
**SIMILAR SPECIES** *Sharp-ribbed Newt (above), Sardinian, and Corsican Brook Newts (p.191), which have different ranges.*

# Sardinian Brook Newt

*Euproctus platycephalus* (Caudata)

**LIVES** *in still and running mountain water; also in riverine woodland, damp scrub, and caves.*

Found only in the mountains of Sardinia, the Sardinian Brook Newt overlaps in distribution with several of the cave-salamanders (p.188). It can be distinguished from these by its preference for more open habitats, and its pointed, instead of stubby, toes. But as with the similarly sluggish cave-salamanders, it has minimal lung function and breathes primarily through its skin.

MALE HIND LEG

**SIZE** *Body up to 8cm; tail up to 7cm.*
**YOUNG** *Lays 50–200 eggs over several months, attached to the underside of stones.*
**DIET** *Invertebrates.*
**STATUS** *Endangered; scarce or rare.*
**SIMILAR SPECIES** *Corsican (below) and Pyrenean (p.190) Brook Newts.*

# Corsican Brook Newt

*Euproctus montanus* (Caudata)

**FOUND** *in and around rivers, streams, and ponds, from sea level to mountains; also in scrub and woodland.*

A sluggish species, the dowdy Corsican Brook Newt is easily distinguished from the only other salamander on the island of Corsica, the vivid Corsican Fire Salamander (p.184). The newt hibernates under stones and logs, in mossy deciduous woodland and scrub and, in hot dry summers, returns to similar areas for aestivation.

**SIZE** *Body up to 7cm; tail up to 6cm.*
**YOUNG** *Lays 20–60 eggs.*
**DIET** *Aquatic invertebrates, slugs, and worms.*
**STATUS** *Locally common.*
**SIMILAR SPECIES** *Sardinian (above) and Pyrenean (p.190) Brook Newts; Luschan's Salamander* (Mertensiella luschani).

# Marbled Newt

*Triturus marmoratus* (Caudata)

**BREEDS** *in still, well-vegetated waterbodies, both permanent and temporary, especially on lowland sandy heaths and in dry, open woodland.*

Almost unmistakable with its emerald green coloration, the Marbled Newt is variably marbled and spotted with black; on land, the green colour is even more intense than in the water. Breeding males have a large, untoothed crest on the back, alternately banded black and whitish; a separate, similar but smaller crest is on the tail, which also has a silvery stripe. Females lack a crest, but have a distinctive orange stripe running down the back to the top margin of the tail. Mainly nocturnal when living on land, taking shelter under rocks or logs, Marbled Newts are active in the day and at night in water during the breeding season. Eggs are laid in spring and summer.

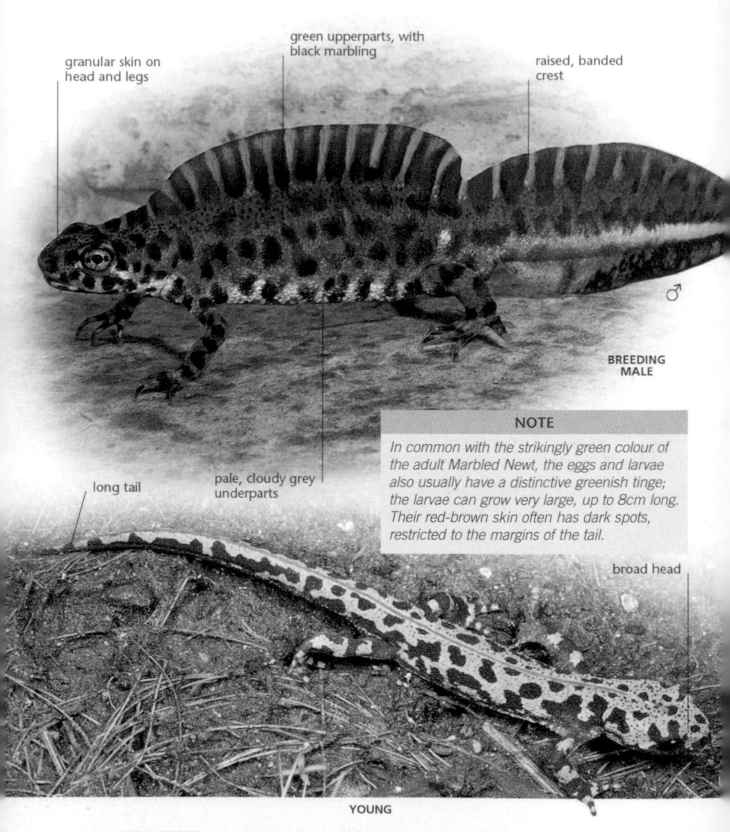


**NOTE**

*In common with the strikingly green colour of the adult Marbled Newt, the eggs and larvae also usually have a distinctive greenish tinge; the larvae can grow very large, up to 8cm long. Their red-brown skin often has dark spots, restricted to the margins of the tail.*

**SIZE** *Body up to 9cm; tail up to 7cm.*
**YOUNG** *Lays 200–400 eggs, singly on aquatic vegetation.*
**DIET** *Aquatic invertebrates, slugs, worms, and insect larvae.*
**STATUS** *Common.*
**SIMILAR SPECIES** *Southern Marbled Newt (p.193), which is smaller, with creamy yellow underparts, and found only in S.W. Iberia; Alpine Newt (p.196).*

# Southern Marbled Newt

*Triturus pygmaeus* (Caudata)

Until recently considered to be the south-western subspecies of the Marbled Newt, the Southern Marbled Newt has a similar green and black pattern to its more widespread relative. However, it is smaller, with distinct spots. Breeding males have a large, smooth, and banded crest.

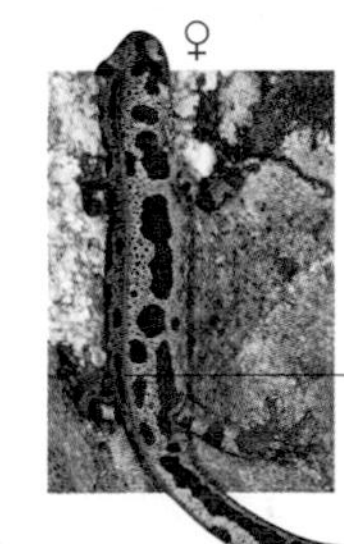


**FOUND** *especially in acidic, standing water, usually within broad-leaved woodland, heathland or on farmland.*

black-spotted, green upperparts

continuous crest from body to tail

♂

creamy yellow, often black-spotted underparts

**SIZE** *Body up to 6cm; tail up to 5cm.*
**YOUNG** *About 200 eggs, laid singly.*
**DIET** *Aquatic invertebrates, slugs, worms, and insect larvae.*
**STATUS** *Near-threatened; locally common.*
**SIMILAR SPECIES** *Marbled Newt (p.192); Alpine Newt (p.196).*

# Italian Crested Newt

*Triturus carnifex* (Caudata)

Formerly treated as a subspecies of the more widespread Great Crested Newt, the Italian Crested Newt replaces it in Italy and the western Balkans. It is shorter than the Great Crested, but more robust, with a smoother skin, and few, if any, white flank spots. Its black spots, above and often below, are rather large and well separated.

deep orange belly

♀

**BREEDS** *in ponds, in deciduous woodland, warm scrubby areas and mountain pastures; takes refuge out of water under logs and stones.*

**SIZE** *Body up to 8cm; tail up to 7cm.*
**YOUNG** *Lays 200-400 eggs.*
**DIET** *Aquatic insects, worms, and snails; small and larval amphibians.*
**STATUS** *Common.*
**SIMILAR SPECIES** *Other Crested Newts (pp.194–195); Alpine Newt (p.196).*

# Great Crested Newt

*Triturus cristatus* (Caudata)

**OCCURS** *in a wide range of weedy, standing waters, usually relatively deep (larvae prefer depths of more than 30cm) and favours sites without predatory fish.*

Also known as the Warty Newt, on account of its granular skin, the Great Crested Newt has blackish skin, especially outside the breeding season. However, when breeding, its ground colour lightens, and black spots and blotches appear, along with white flecks on the head and flanks. The distinctive, jagged crest, with a gap at the base of the tail, and silvery tail flash feature only in males. Largely nocturnal, they also hibernate, sometimes buried in mud, but usually under logs and stones, up to 200m from the breeding site.

JUVENILE

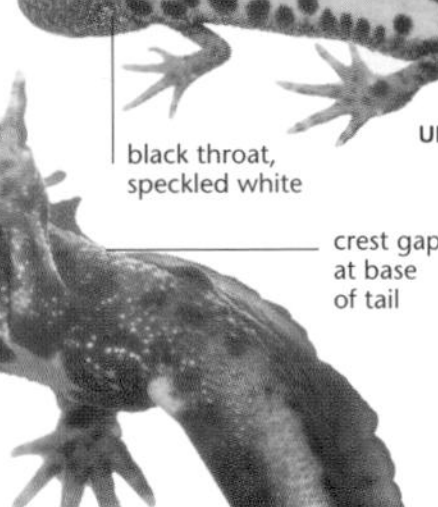


UNDERSIDE

BREEDING MALE

**NOTE**

*As with many newts, Great Crested Newts are most easily spotted during the breeding season by torchlight at night, when they may be seen rising to the surface; however, small populations are perhaps best located by searching water plants for their eggs.*

orange lower tail margin

dark, almost black, skin

granular ("warty") skin

♀

**SIZE** *Body up to 8cm; tail up to 7.5cm.*
**YOUNG** *Lays 200–400 eggs over many months; hatch in three weeks.*
**DIET** *Aquatic insects, worms, and snails; small and larval amphibians.*
**STATUS** *Common, but declining.*
**SIMILAR SPECIES** *Italian Crested Newt (p.193), which has large, round black belly spots; Balkan Crested Newt (p.195), which has a pale throat; Danube Crested Newt (p.195), which has irregular black blotching.*

# Balkan Crested Newt

*Triturus karelinii* (Caudata)

The largest of the Crested Newt group, the Balkan Crested Newt is often found at relatively high altitudes, although only in the eastern Balkans. It may be distinguished from the other newts in the group by its pale throat, sometimes with darker spots, and the usually small, uniform size of the dark spots on its orange belly. The crest of the breeding male is generally smooth in profile, and not fully indented at the base of the tail.

**LIVES** *in ponds, deciduous woodland, warm scrubby areas, and often at montane levels; takes refuge out of the water under logs and stones.*

pale throat
fairly smooth skin
bluish sheen
tail and back crests not fully separate
♂

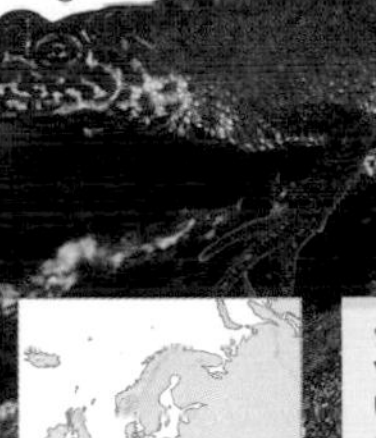

**SIZE** *Body up to 8cm; tail up to 8cm.*
**YOUNG** *Lays 200–400 eggs.*
**DIET** *Aquatic insects, worms, and snails; small and larval amphibians.*
**STATUS** *Locally common.*
**SIMILAR SPECIES** *Some other Crested Newts (pp.193–195), which have black throats.*

# Danube Crested Newt

*Triturus dobrogicus* (Caudata)

The smallest and rarest of the Crested Newts, the Danube Crested Newt has an interesting distribution, centred on the floodplain and delta of the Danube river system. With its coarsely granular skin, it resembles the Great Crested, but the breeding male has an even taller spiky crest, and it lacks white stippling on the flanks. Its belly spots are often enlarged, forming irregular blotches, sometimes obscuring much of the deep orange belly.

black throat
more black than orange on belly

UNDERSIDE

**FAVOURS** *still and slow-moving lowland waterbodies, usually well vegetated; also in riparian woodland.*

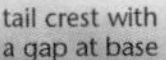


very coarse skin
red-brown upperparts with black spots
♂
small head

**SIZE** *Body up to 7cm; tail up to 6cm.*
**YOUNG** *Lays probably 200–400 eggs.*
**DIET** *Aquatic insects, worms, and snails; small and larval amphibians.*
**STATUS** *Near-threatened; scarce.*
**SIMILAR SPECIES** *Other Crested Newts (pp.193–195).*

# Alpine Newt

*Triturus alpestris* (Caudata)

**INHABITS** *still and slow-moving water, from shady woodland pools in lowland, to clear mountain ponds, often lacking significant vegetation.*

A beautiful newt in its breeding colours, the male Alpine Newt has dark grey-blue upperparts, which often show a deep purple sheen. Breeding males have a smooth yellow and black crest, while female Alpine Newts are more of an olive grey colour. Both show black spots, especially on the pale flanks, and deep orange or red, unspotted underparts (although the throat may have dark spots). The orange extends along the underside of the tail in the female. In spite of its name, it is not restricted to upland areas, except in the south of its range. However, it can tolerate cold and survive under ice. Its hibernation sites are usually cool, shady places very close to water.

**NOTE**

*Alpine Newts show much variation across their range: some have red bellies, others have spotted throats, while certain Balkan mountain species are wholly or partly neotenous, retaining feathery larval gills on an adult body.*

**SIZE** *Body up to 6cm; tail up to 6cm.*
**YOUNG** *Lays 250–500 eggs in early spring, hatching within a month.*
**DIET** *Aquatic invertebrates, tadpoles, and worms, mostly from the mud in their ponds.*
**STATUS** *Common.*
**SIMILAR SPECIES** *Marbled (p.192) and Crested Newts (pp.193–196), outside the breeding season; Montadon's and Bosca's Newts (p.197).*

# Montadon's Newt

*Triturus montadoni* (Caudata)

Largely montane and rather small, the pale, yellow-brown Montadon's Newt is variably (but usually weakly) sprinkled with black specks. The belly is yellow or pale orange, and the throat is paler still; both are unspotted. Ridges running down the back, either side of the spine, give a distinctively square cross-section to the body.

**FREQUENTS** *upland ponds and streams, within coniferous woodland and pastures; tolerant of pollution.*

black-mottled sides

♂

three grooves on head

plain face

**SIZE** *Body up to 5cm; tail up to 5cm.*
**YOUNG** *Lays 50–250 eggs.*
**DIET** *Invertebrates.*
**STATUS** *Locally common.*
**SIMILAR SPECIES** *Alpine Newt (p.196), which is darker above and redder below; Smooth Newt (p.198), which is spotted below.*

# Bosca's Newt

*Triturus boscai* (Caudata)

The only small newt over most of its range, in the western part of the Iberian Peninsular, Bosca's Newt is unmistakable, except where it overlaps with Palmate and Alpine Newts. It is strongly aquatic, some individuals remaining in the water year-round. When they do venture onto land, the smooth skin becomes granular, and the parotid glands and mid-line of the back often turn reddish.

**OCCUPIES** *shallow, clean pools, streams, and artificial water bodies, often with little vegetation.*

orange below, edged white with black spots

**SIZE** *Body up to 5cm; tail up to 5cm.*
**YOUNG** *Lays 100–250 eggs.*
**DIET** *Invertebrates.*
**STATUS** *Near-threatened; locally common, some populations declining.*
**SIMILAR SPECIES** *Alpine Newt (p.196); Palmate Newt (p.199).*

# Smooth Newt

*Triturus vulgaris* (Caudata)

**LIVES** *on land in damp woodland, marshes, gardens, and other habitats; breeding in ponds, ditches, lake margins, and slow-moving rivers.*

One of the most widespread and abundant amphibians in Europe, it is not surprising that an alternative name for the Smooth Newt is the Common Newt. It is much more terrestrial than many other newts, typically living in water as an adult only for breeding, between March and July. Non-breeding animals are olive brown, with a black-spotted throat and two ridges on the head which join near the snout. They have darker spots and a black-spotted, orange belly. Largely nocturnal, they shelter (as well as hibernate) under logs, stones, or leaf-litter. Breeding males, in contrast, are very showy, with a wavy crest up to 1cm high, and clear black face stripes.

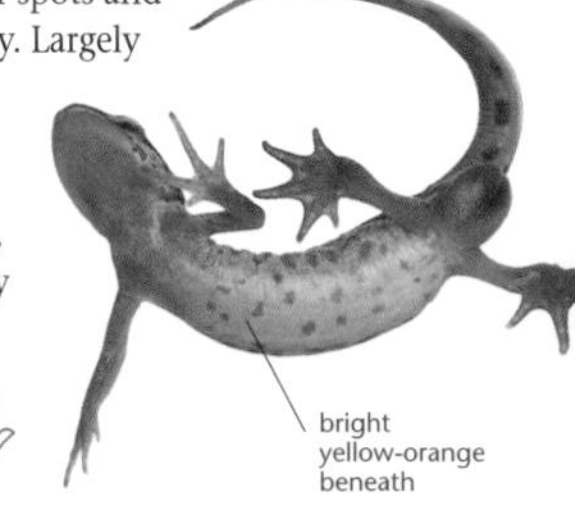


**NOTE**

*Sometimes mistaken for the Great Crested Newt on account of its equally flamboyant crest, the most obvious point of distinction is that the crest of the breeding male Smooth Newt is continuous from the back onto the tail.*

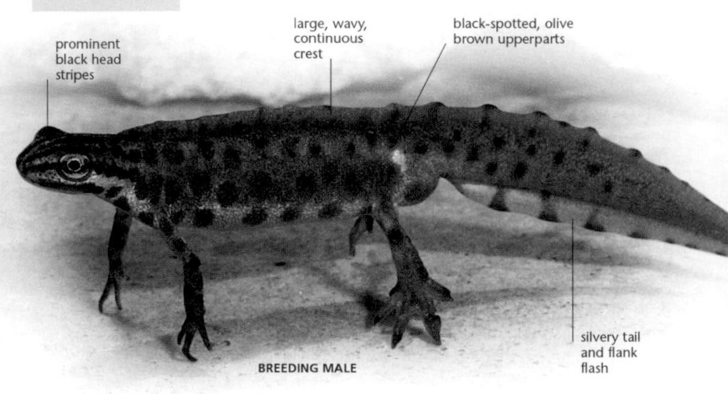


**BREEDING MALE**

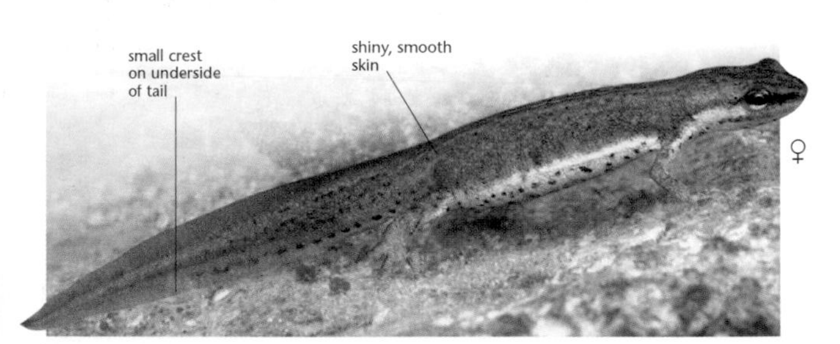


**SIZE** *Body up to 6cm; tail up to 5cm.*
**YOUNG** *Lays 200–500 eggs; hatch in 2–3 weeks.*
**DIET** *Insects, worms, and other invertebrates; tadpoles.*
**STATUS** *Common.*
**SIMILAR SPECIES** *Montadon's (p.197) and Italian (p.199) Newts, which are generally paler, less spotted below, and without dark head stripes; Palmate Newt (p.199), which has a whitish, unspotted throat.*

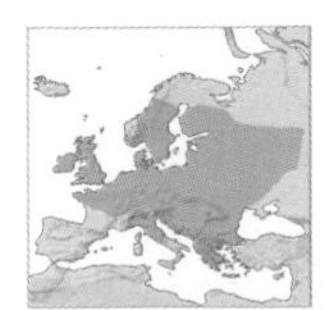

# Palmate Newt

*Triturus helveticus* (Caudata)

One of the smaller European species, the Palmate Newt gets its name from the black webs which develop on the hind feet of the male during the breeding season; its breeding appearance is also characterized by a prominent crest on the tail, which terminates in a distinct filament. At all times, Palmate Newts have a blackish stripe through the eye, a pale yellow belly, and unspotted throat. Most adults leave the breeding pools in late summer and hibernate under rocks and logs.

**FOUND** *in shallow, still, often acidic waterbodies, usually well vegetated, in woodland, heathland, farmland, and mountains; can tolerate brackish conditions.*

**SIZE** *Body up to 4.5cm; tail up to 4.5cm.*
**YOUNG** *Lays 300–400 eggs.*
**DIET** *Invertebrates, worms, and tadpoles.*
**STATUS** *Common.*
**SIMILAR SPECIES** *Bosca's Newt (p.197), which is darker below; female Smooth Newt (p.198), which is more spotted below.*

# Italian Newt

*Triturus italicus* (Caudata)

Europe's smallest newt, the Italian Newt is restricted to the southern part of the Italian peninsula. The olive brown upperparts are spotted with black, while the belly is yellow-orange, with small, scattered, dark spots. Uniquely, the throat is a darker orange than the belly, and usually almost unspotted. In the breeding season, the male's tail develops a smooth, low crest, and a projecting filament at the end.

**INHABITS** *ponds, temporary pools, and artificial waterbodies, generally in the lowlands, in scrub, cultivated areas, and open woodland.*

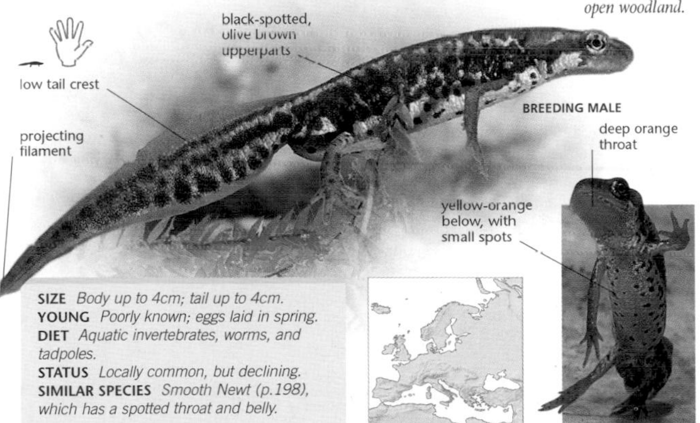


**SIZE** *Body up to 4cm; tail up to 4cm.*
**YOUNG** *Poorly known; eggs laid in spring.*
**DIET** *Aquatic invertebrates, worms, and tadpoles.*
**STATUS** *Locally common, but declining.*
**SIMILAR SPECIES** *Smooth Newt (p.198), which has a spotted throat and belly.*

# Yellow-bellied Toad

*Bombina variegata* (Anura)

**OCCURS** *in shallow puddles, ponds, and small streams, in marshes, woodland, and on mountains.*

A very small toad, the Yellow-bellied Toad is unremarkable, until its brightly coloured underside is seen. If threatened, it arches its back and raises its legs to display its colours as a warning of its toxic skin secretions. A sociable species, it is seen floating on water, legs outstretched, the males calling a soft, musical "poop", especially in spring, after hibernation.

UNDERSIDE

**SIZE** *4–5cm.*
**YOUNG** *Lays 100–150 eggs; May–July, attached to stones or vegetation.*
**DIET** *Aquatic insects and other invertebrates.*
**STATUS** *Common.*
**SIMILAR SPECIES** *Fire-bellied Toad (below); Italian forms sometimes called* B. pachypus.

# Fire-bellied Toad

*Bombina bombina* (Anura)

**FAVOURS** *lowland, still water ponds, lakes, and ditches, usually with clean water and dense vegetation.*

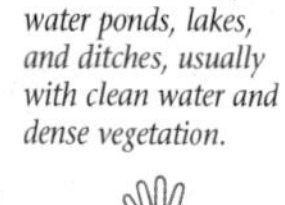

Similar in size and shape to the Yellow-bellied Toad (and hybridizing where they occur together), the Fire-bellied Toad is most easily distinguished by its orange-red belly markings, the dark patches bearing numerous white spots. Many of the warts on the back are darker than the ground colour and also more rounded. Hibernation sites are either under rocks or logs, or buried in damp, soft mud.

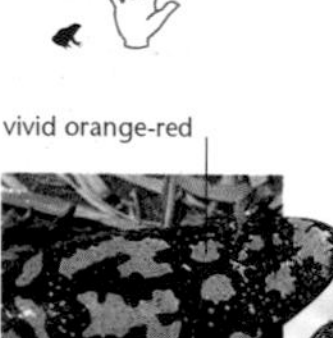


UNDERSIDE

**SIZE** *4–5cm.*
**YOUNG** *Lays about 300 eggs; singly or in small batches; attached to submerged plants.*
**DIET** *Invertebrates, especially insects.*
**STATUS** *Locally common.*
**SIMILAR SPECIES** *Yellow-bellied Toad (above), which has a paler underside.*

# Common Midwife Toad

*Alytes obstetricans* (Anura)

Named for the male's spawn-carrying habits, the Common Midwife Toad is a small, plump toad, with prominent eyes that have golden irises and vertical pupils. Its upper skin surface is grey-brown (although often olive or green, especially in some Iberian forms), with a sparse covering of small warts. Its hind foot webs extend to half the length of the toes. A Midwife Toad has a rather mixed gait, running like a typical toad, but also jumping like a frog. Listen for the high-pitched "poo...poo...poo" calls at night of this secretive and largely nocturnal species.

**LIVES** *in grassland and scrub, open woodland, gardens, and parks, with access to breeding ponds; hides by day under logs or rocks – or in burrows in soft sand, which are also favoured hibernation sites.*

**NOTE**

*Male Midwife Toads twine fertilized strings of eggs around their midriff, remaining in damp areas to prevent them from drying, until they hatch after about eight weeks. The tadpoles are then released into ponds and ditches.*

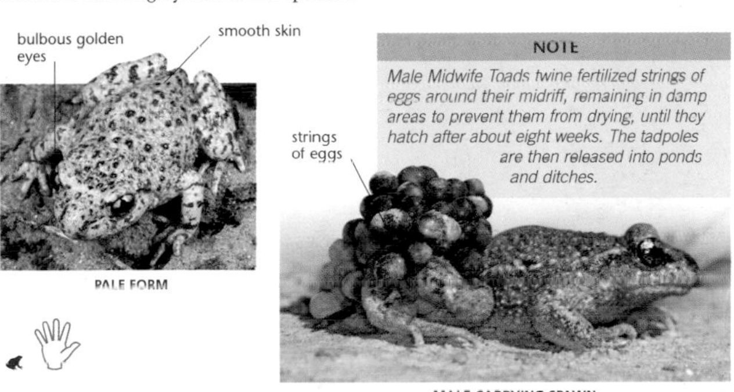


PALE FORM

MALE CARRYING SPAWN

**SIZE** *Up to 5cm.*
**YOUNG** *Lays 2–4 clutches of up to 80 eggs; May–August.*
**DIET** *Insects, spiders, and worms.*
**STATUS** *Common.*
**SIMILAR SPECIES** *Parsley Frog (p.208); Iberian Midwife Toad* (A. cisternasii), *which is larger; Southern Midwife Toad* (A. dickhilleni); *Mallorcan Midwife Toad* (A. muletensis), *which is smaller.*

# Common Spadefoot

*Pelobates fuscus* (Anura)

**OCCUPIES** *lowland sandy areas, heaths, cultivated fields and open woodland, with vegetated ponds and ditches for breeding.*

Also known as the Garlic Toad, on account of the smell it produces when threatened, the Common Spadefoot does indeed have a spade on its foot, a pale swelling under the hind foot which increases its efficiency as a digger. It is capable of digging burrows up to a metre deep, which then collapse, so that the toad has to push its way out, using its armoured skull. In such burrows it will hibernate in winter, and aestivate in times of drought; normal daylight refuges are shallower. The smooth skin of the Common Spadefoot is variably marked and marbled with brown and green.

toad in daytime refuge

BURROW

PALE FORM

**NOTE**

*The development of the Spadefoot tadpole is slow, typically over a period of 4–5 months, resulting in the growth of tadpoles to a remarkably large size (up to 16cm), although after metamorphosis, even big tadpoles become juvenile frogs of only 2–3cm length.*

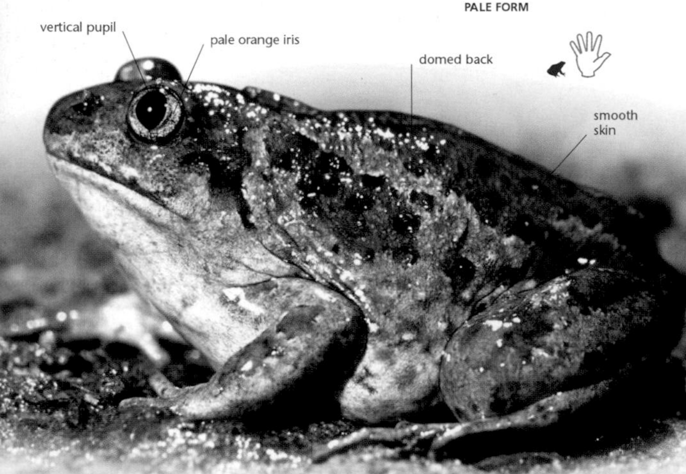


**SIZE** *Body 6–8cm.*
**YOUNG** *Lays 1,000–3,000 eggs in May in broad bands around aquatic plants; hatch within 10 days.*
**DIET** *Insects, snails, and worms.*
**STATUS** *Common.*
**SIMILAR SPECIES** *Western and Eastern Spadefoots (p.203); other Toads (pp.204–206), which have horizontal pupils and no spade.*

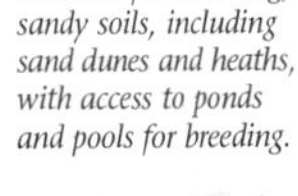

# Western Spadefoot

*Pelobates cultripes* (Anura)

A little larger than the Common Spadefoot, the Western Spadefoot is more marbled with green; it also lacks a lump on its head. It has a greater tendency to breed in temporary pools: this is a risky strategy as any unmetamorphosed larvae are killed if the pools dry out. But there are potential benefits too: temporary waterbodies are less likely to contain predators of the larvae, and the adults further enhance their reproductive chances by laying a very large number of eggs.

**FAVOURS** *free-draining, sandy soils, including sand dunes and heaths, with access to ponds and pools for breeding.*

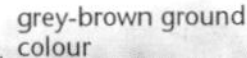

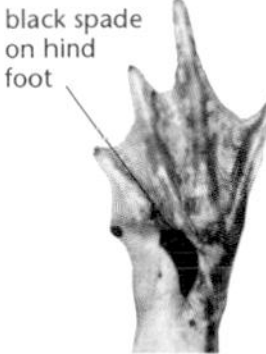

HIND FOOT

**SIZE** *Body 7–10cm.*
**YOUNG** *Lays up to 7,000 eggs.*
**DIET** *Insects, snails and worms.*
**STATUS** *Common.*
**SIMILAR SPECIES** *Eastern (below) and Common (p.202) Spadefoots, which have pale spades; Toads (p.204–205).*

# Eastern Spadefoot

*Pelobates syriacus* (Anura)

A species of south-eastern Europe, the Eastern Spadefoot is very similar to the Common Spadefoot, with which it just overlaps in range. They both have a pale spade, but Eastern lacks a lump on the back of its head and has deeply indented webbing between the hind toes. It also has green-brown blotches, sometimes dark-edged, on its back. Like all spadefoots, it is largely nocturnal, and may both hibernate and aestivate in soft sand.

**OCCURS** *in cultivated land, open scrub and woodland, and sandy areas, near pools.*

**SIZE** *Body up to 9cm.*
**YOUNG** *Lays 2,000 to 4,000 eggs.*
**DIET** *Insects, snails, and worms.*
**STATUS** *Common.*
**SIMILAR SPECIES** *Western Spadefoot (above); Common Spadefoot (p.202); Toads (p.204–205); Green Toad (p.206).*

# Common Toad

*Bufo bufo* (Anura)

**FOUND** *in a wide range of habitats, such as marshes, woodland, heathland, gardens, and mountain pastures, with access to pools and ponds for breeding.*

A very widespread and robust species, the Common Toad is generally uniform brown in colour, its skin covered in numerous warts. Across its range, it varies primarily in size – females are bigger than males, while southern toads are the largest and tend to have more horny warts. The parotid glands, behind the eyes, are prominent and, when viewed from above, are divergent, rather than parallel as in other members of its family. This gland exudes a secretion that repels predators. Common Toads are mainly nocturnal, except when breeding, and hibernate under logs, rocks, tree roots, or in leaf mould. Their normal gait is a lumbering walk, although they will hop if alarmed.

**NOTE**

*Following their emergence from hibernation, these toads migrate en masse to breeding ponds. Where such routes intersect with roads, hundreds are often run over by passing traffic.*

**SIZE** *Body 8–15cm, especially large in the south of the range.*
**YOUNG** *Lays 1,000–8,000 eggs; hatching in three weeks.*
**DIET** *Slugs, insects, spiders, and worms; large toads feed on other amphibians, small reptiles, and small mammals.*
**STATUS** *Common.*
**SIMILAR SPECIES** *Spadefoots (pp.202–203); Natterjack Toad (p.205), which has a pale stripe down its back; Green Toad (p.206).*

# Natterjack Toad

*Bufo calamita* (Anura)

A small but robust toad, the Natterjack Toad is widespread across western Europe, although it has very specific habitat preferences in the north of its range. Here it frequents sandy areas, where it can burrow for daytime refuge, hibernation, and aestivation, in sites which are used often communally. Its characteristic marking is a yellow stripe down its spine, sometimes appearing very bright (although occasionally absent) and it has parallel parotid glands that are often orange. Its gait is very distinctive – instead of walking like most toads, the Natterjack has a scurrying run, rather like a mouse.

**PREFERS** *sandy heaths and dunes, with ponds, often temporary, for breeding; also marshes, salt marshes, and even mountains in the south.*

eggs in long, single strings

**SPAWN**

**NOTE**

*A very noisy species when breeding, males have a vocal sac under the chin, which amplifies its repeated, rolling croak; a chorus of males can be heard at a distance of several kilometres, and may be mistaken for the churring of a Nightjar.*

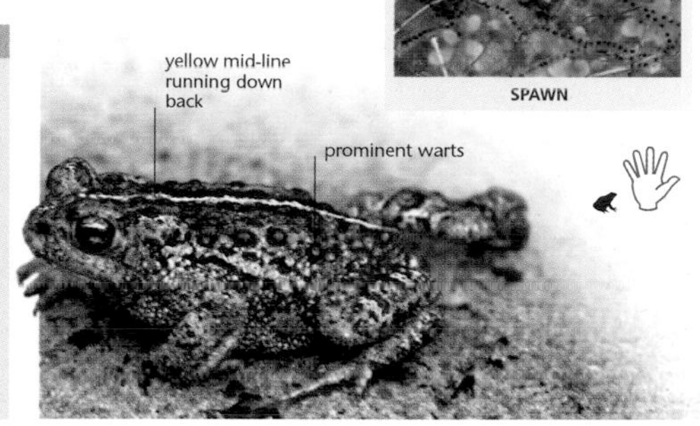

**SIZE** *Body 6–8cm.*
**YOUNG** *Lays 2,000–5,000 eggs, hatching rapidly; may develop very quickly in shallow water. Tadpoles small and black, with bronze spots.*
**DIET** *Slugs, spiders, insects and worms.*
**STATUS** *Common, but declining in some peripheral parts of its range.*
**SIMILAR SPECIES** *Spadefoots (pp.202–203); Common Toad (p.204); Green Toad (p.206), which has a well-defined, greenish pattern.*

# Green Toad

*Bufo viridis* (Anura)

**INHABITS** *sandy and coastal areas, and marshland, especially at low altitudes; often seen in gardens and villages.*

The Green Toad is an attractively marked species, its green pattern, more distinctive in females, set off against a pale olive grey background. It has parallel parotid glands and horizontal pupils. Largely nocturnal, breeding males have a continuous trilling call, which can be mistaken for the song of a Mole-cricket.

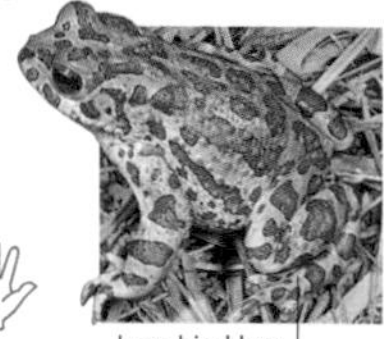


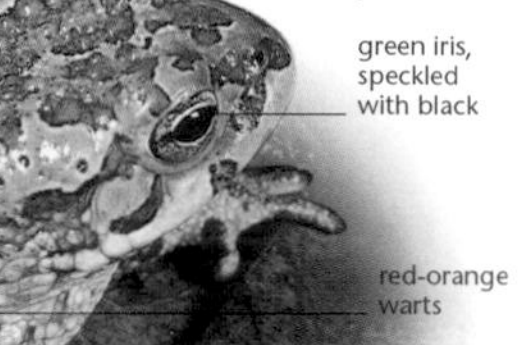

**SIZE** *6–10cm.*
**YOUNG** *Up to 12,000 eggs: in rows of two to four, in long strands.*
**DIET** *Insects, slugs, spiders, and worms.*
**STATUS** *Common.*
**SIMILAR SPECIES** *Eastern Spadefoot (p.203); Toads (p.205–206).*

# Stripeless Tree-frog

*Hyla meridionalis* (Anura)

**FOUND** *in and around marshes, reedbeds, scrubby riverbanks, garden ponds, and also in banana plantations.*

In common with all tree-frogs, the Stripeless Tree-frog spends much of its time singing from tall vegetation in damp areas; its song is a deep, resonant croak, repeated every second or more, often as part of a large chorus. Its skin is generally green, but can change to blue, brown, or yellow to match its surroundings. This frog is mostly nocturnal.

**SIZE** *3–6.5cm.*
**YOUNG** *Up to 60 clumps of 10–30 eggs laid among aquatic vegetation.*
**DIET** *Insects and other invertebrates.*
**STATUS** *Common.*
**SIMILAR SPECIES** *Common Tree-frog (p.207), which has an extended eye-stripe.*

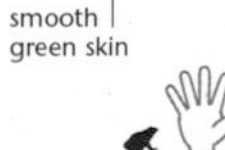

# Common Tree-frog

*Hyla arborea* (Anura)

Although typically bright green, the skin of the Common Tree-frog can change rapidly from yellow to almost black, according to its background, age, state of skin shedding, and light conditions. It is an agile species, climbing well with the aid of suction pads on its toes, and often leaping between the branches of trees to catch insects in flight. Tree-frogs hibernate in holes in the ground or under tree roots and, except at the peak of the breeding season, are largely nocturnal, although they often sunbathe on leaves.

**FAVOURS** *reedbeds, marshland, riverbanks, and other areas with tall vegetation close to water, including gardens.*

suction pads at tips of toes

dark stripe from nostrils to flanks

long limbs

very smooth, bright green skin

GREY FORM

**NOTE**

*At night during the breeding season, males are very vocal, producing a rapid, quacking croak; en masse their croaks merge into a continuous cacophony, which can be heard over several kilometres.*

whitish, granular underparts

yellowish vocal sac under chin

BREEDING MALE

**SIZE** *3–5cm.*
**YOUNG** *Up to 1,000 eggs produced in clumps of up to 60, hatching in about three weeks; tadpoles small, golden-olive, with high tail crests.*
**DIET** *Insects and other invertebrates.*
**STATUS** *Common.*
**SIMILAR SPECIES** *Stripeless Tree-frog (p.206); Italian Tree-frog* (H. intermedia); *and Tyrrhenian Tree-frog* (H. sarda).

# Painted Frog

*Discoglossus pictus* (Anura)

**OCCURS** *around cultivated land, scrub, open woodland, and marshland, always close to water.*

Variably coloured, most often in shades of olive and brown, the Painted Frog may be plain, blotched, or striped; some individuals have one or three noticeable pale back stripes. It can be identified by its rounded (rather than horizontal) pupil and very inconspicuous ear-drum. Another distinguishing feature is its disc-shaped tongue, which does not protrude, meaning its prey has to be captured in its jaws.

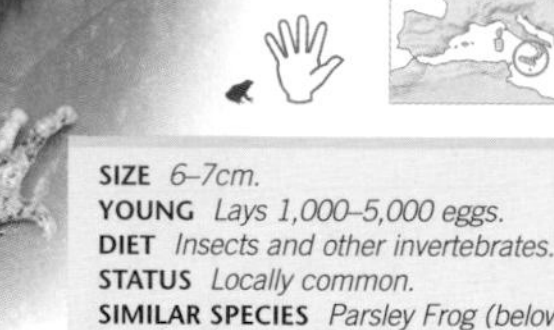


**SIZE** *6–7cm.*
**YOUNG** *Lays 1,000–5,000 eggs.*
**DIET** *Insects and other invertebrates.*
**STATUS** *Locally common.*
**SIMILAR SPECIES** *Parsley Frog (below); other close relatives include* D. galganoi, D. jeanneae, D. sardus *and* D. montalentii.

# Parsley Frog

*Pelodytes punctatus* (Anura)

**BREEDS** *in open but well-vegetated ponds, pools, and ditches; out of water in damp shady areas, taking refuge by day under stones, in burrows, or in rock crevices.*

Generally a brown or olive colour, the Parsley Frog is capable of rapid colour changes; the warts are often darker on the back and more orange on the flanks. Generally nocturnal, it is also very agile, often climbing rocks and walls. It burrows in soft sand.

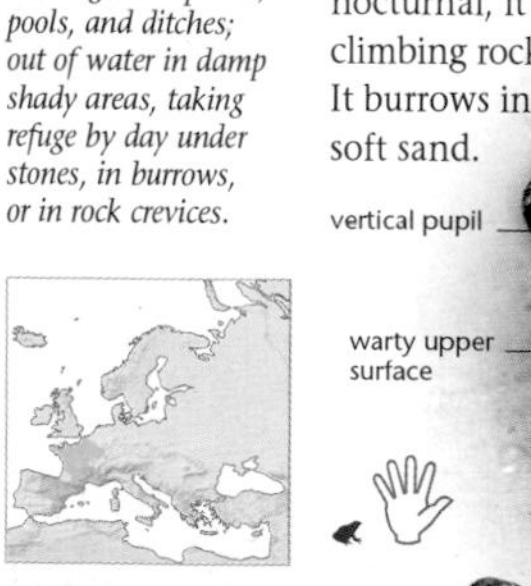

**SIZE** *4–5cm.*
**YOUNG** *Up to 1,600 eggs.*
**DIET** *Insects and other invertebrates.*
**STATUS** *Common.*
**SIMILAR SPECIES** *Painted Frog (above); Common Midwife Toad (p.201); Iberian Parsley Frog* (P. ibericus).

# Common Frog

*Rana temporaria* (Anura)

The most widespread of the brown frog group, the Common Frog is relatively robust, with a wide, blunt snout and two parallel ridges running down the back. Its colour is variable but the ground colour is usually yellowish to olive brown, and the markings dark brown to almost black. The colour often intensifies during the breeding season, when the female develops a granular skin and the male's throat often turns bluish. Active during the day and the night, the northern populations hibernate under logs and stones, or in leaf mould or burrows.

**FOUND** *in lowland meadows, marshes, gardens, and mountain pastures, with access to shallow ponds and ditches for breeding.*

large ear-drum

fully webbed hind feet

PALER FORM

HIND FOOT

**SIZE** *6–8cm.*
**YOUNG** *Lays 1,000 to 4,000 eggs.*
**DIET** *Slugs, snails, insects, worms, and other invertebrates.*
**STATUS** *Common.*
**SIMILAR SPECIES** *All brown frogs (p.210–213); especially Moor Frog (p.211).*

**NOTE**

*When frogs gather at traditional breeding pools, the males utter a low, purring croak and grab anything that moves; if it happens to be a female, she can attract a whole host of suitors, and in the ensuing melée may even be drowned.*

# Iberian Frog

*Rana iberica* (Anura)

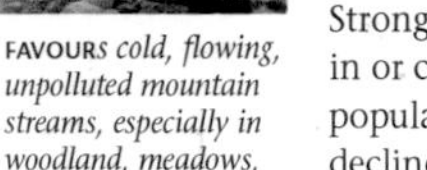

**FAVOURS** *cold, flowing, unpolluted mountain streams, especially in woodland, meadows, or scrub, often well above the tree-line.*

Strongly aquatic, the Iberian Frog is almost always found in or close to water within its isolated mountain populations, some of which are considered to be in serious decline. Its long legs give it the agility to climb rocks, while its fully webbed hind feet provide the power to swim in fast-flowing streams. Its throat is mottled, with a pale central line.

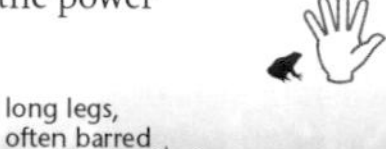

HIND FOOT

long legs, often barred

often red-brown above

dark eye mask

fully webbed

small ear-drum

**SIZE** *Up to 7cm.*
**YOUNG** *Up to 500 eggs; November–April.*
**DIET** *Insects and other invertebrates.*
**STATUS** *Near-threatened; locally common.*
**SIMILAR SPECIES** *Common Frog (p.209), which has shorter legs; Agile Frog (p.212), which has longer legs and a large ear-drum.*

# Pyrenean Frog

*Rana pyrenaica* (Anura)

**OCCURS** *in rocky mountain rivers that are fast-flowing and well oxygenated.*

The smallest European brown frog, the Pyrenean Frog is unique to a small area of the Pyrenees, where it is found in relatively high-altitude streams. However, it should be noted, the Common Frog also occurs at similar and greater heights. The delicate build, usually weak markings, and widely separated nostrils, greater than the distance between eye-bulges, are key identifying features.

pale to reddish colour

whitish throat, usually spotted

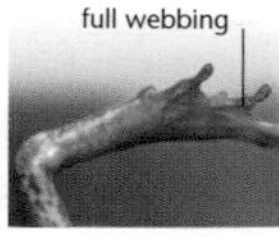

full webbing

HIND FOOT

**SIZE** *Up to 5cm.*
**YOUNG** *Up to 150 very large eggs, laid under stones.*
**DIET** *Insects and other invertebrates.*
**STATUS** *Vulnerable; locally common.*
**SIMILAR SPECIES** *Common Frog (p.209), which is larger, with a more pointed snout.*

# Moor Frog

*Rana arvalis* (Anura)

In comparison with the Common Frog, the member of the brown frog group with which it overlaps most extensively, the Moor Frog is more nocturnal, and has a sharply pointed snout. It also has shorter legs, except in the south of its range, where long-legged forms can be confused with Agile Frogs. Its pattern and colours are very variable, but the dark spots often form broken stripes. There is often a broad, pale brown stripe down the middle of the back, sometimes with narrower stripes on either side. Moor Frogs usually hibernate in soft mud at the edge of a breeding pond.

**INHABITS** *flood meadows, peat bogs, and damp woodland, near breeding pools and ponds, in lowland as well as hilly areas.*

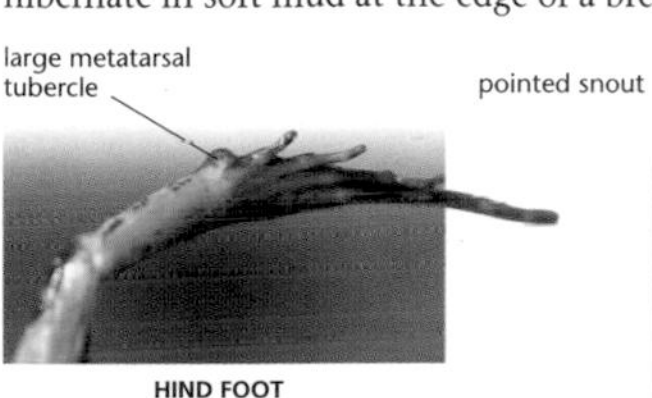

**HIND FOOT**

**DARKER FORM**

**NOTE**

*Rather obscure characteristics are important in the definitive identification of many frogs. The metatarsal tubercle is a swelling at the base of the shortest toe on the hind foot: It is small and soft in the Common Frog, but large and horny in the Moor Frog.*

**SIZE** *6–8cm.*
**YOUNG** *Lays 2,000–4,000 eggs in 1–2 masses, which sink to the bottom of the pond and hatch in 2–4 weeks.*
**DIET** *Insects, worms, and other invertebrates.*
**STATUS** *Common.*
**SIMILAR SPECIES** *Common Frog (p.209), which has a more rounded snout and a much smaller metatarsal tubercle; Agile Frog (p.212).*

# Agile Frog

*Rana dalmatina* (Anura)

**LIVES** *in open, moist broad-leaved woodland and lowland marshes, and meadows.*

This elegantly proportioned species has a slender, flattened body, pointed snout, and long, slim legs. It is delicately coloured, with a yellow- to pink-tinged pale brown ground colour and very sparse dark spotting. The tadpoles are also distinctive – pale rufous brown above and white below, with gold spots. A poor swimmer, the Agile Frog is very much at home on land; females tend to hibernate under logs, but males usually return to the water and hibernate in the mud at the bottom.

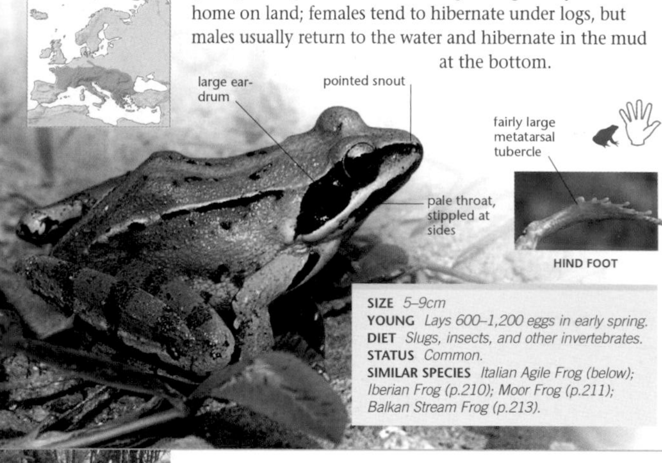


**SIZE** *5–9cm*
**YOUNG** *Lays 600–1,200 eggs in early spring.*
**DIET** *Slugs, insects, and other invertebrates.*
**STATUS** *Common.*
**SIMILAR SPECIES** *Italian Agile Frog (below); Iberian Frog (p.210); Moor Frog (p.211); Balkan Stream Frog (p.213).*

# Italian Agile Frog

*Rana latastei* (Anura)

**FAVOURS** *lowland riverine woodland, poplar plantations, and similar humid, shady habitats, with access to ponds, ditches, and streams for breeding.*

Very similar to the Agile Frog, the Italian Agile Frog's snout is rather less pointed. It also tends to be a darker colour, often with more obvious dark spots and blotches, especially on the legs. Breeding males, in particular, also have a distinctive reddish wash on the underparts and legs, instead of the yellowish colour found in the Agile Frog and the Stream Frogs.

**SIZE** *Up to 7.5cm.*
**YOUNG** *300–400 eggs.*
**DIET** *Insects and other invertebrates.*
**STATUS** *Vulnerable; scarce and declining.*
**SIMILAR SPECIES** *Agile Frog (above); Balkan Stream Frog (p.213); Italian Stream Frog* (R. italica), *which has nostrils wider apart.*

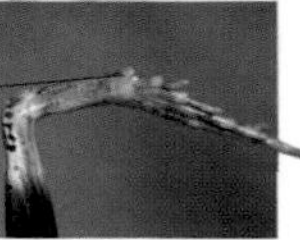

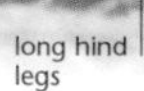

# Balkan Stream Frog

*Rana graeca* (Anura)

Distinguished from other European brown frogs by the spacing of its nostrils, those of the Balkan Stream Frog are further apart than the distance between the nostril and the eye. It has a variable brown colour, often with sparse dark blotches, sometimes forming a V-shape between the shoulders. It typically basks on the edge of water and hibernates generally on land.

**FOUND** *mainly in and around cool, rocky mountain streams and pools, and in caves.*

flattened body

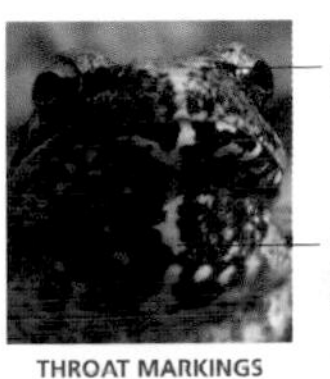


THROAT MARKINGS

metatarsal tubercle

HIND FOOT

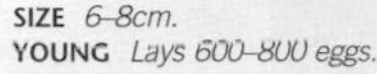

**SIZE** *6–8cm.*
**YOUNG** *Lays 600–800 eggs.*
**DIET** *Insects, worms, and slugs.*
**STATUS** *Scarce.*
**SIMILAR SPECIES** *Agile and Italian Agile Frogs (p.212), which are more slender; Italian Stream Frog* (R. italica).

# American Bullfrog

*Rana catesbeiana* (Anura)

The introduced American Bullfrog is named for its loud, resonant groaning call, produced by males on warm nights in the breeding season. It is much larger than native European frog species. Its tadpoles are also large, growing to 16cm in length; they are olive green above and pale below, with ridges on the back.

**INHABITS** *lakes, pools, marshes, and ditches, usually with luxuriant vegetation.*

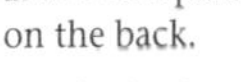


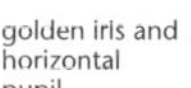


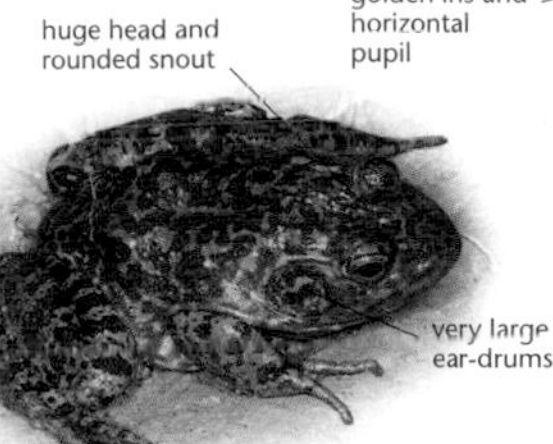


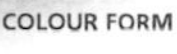

COLOUR FORM

**SIZE** *12–20cm.*
**YOUNG** *Up to 20,000 eggs.*
**DIET** *Insects, crustaceans, fish, amphibians, waterbirds' chicks, voles, and mice.*
**STATUS** *Scarce but increasing.*
**SIMILAR SPECIES** *Oversized water frogs (pp.214–216), which have ridges on the back.*

# Edible Frog

*Rana* kl. *esculenta* (Anura)

The insertion of *kl.* into the scientific name of the Edible Frog is important: it is short for the Greek *klepton* and signifies that the "species" in question is derived from hybridization between two true species. In this case, the parental types are Marsh and Pool Frogs; the Edible Frog is in many respects intermediate between the two, but variable and sometimes more like one than the other, depending on the genetic contribution of each parental type.

**FREQUENTS** *lakes, ponds, and marshes, including brackish water and hilly areas; hibernates in water, or under logs and stones on land.*

large metatarsal tubercle

**REAR LEG**

**SIZE** *7–12cm.*
**YOUNG** *Up to 10,000 eggs.*
**DIET** *Invertebrates, fish, and amphibians.*
**STATUS** *Common.*
**SIMILAR SPECIES** *Pool Frog (below) and Marsh Frog (p.216), which differ in the size of the metatarsal tubercle.*

# Pool Frog

*Rana lessonae* (Anura)

Very similar to other water frogs, especially Edible Frog of which it is a parental form, the Pool Frog is variable in both colour and pattern: one of its most constant features is the very large hemispherical metatarsal tubercles, more than half the length of the first toe. It also has a loud "quacking" croak, rather than the frantic repetition of the Marsh Frog.

**FAVOURS** *marshland, meadows, heaths, and wet woodland, with shallow pools for breeding; sometimes with Marsh Frogs in larger waterbodies.*

large metatarsal tubercle

**HIND LEG**

**SIZE** *6–8cm.*
**YOUNG** *Up to 3,000 eggs.*
**DIET** *Invertebrates, fish, and amphibians.*
**STATUS** *Common; extinct in Britain.*
**SIMILAR SPECIES** *Edible (above) and Marsh Frogs (p.216); Italian Pool Frog* (R. bergeri) *and Italian Hybrid Frog* (R. *kl.* hispanica).

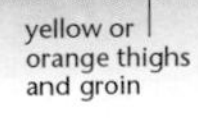

# Iberian Water Frog

*Rana perezi* (Anura)

The Iberian Water Frog is found in southwest Europe and is the only species of frog in the Balearics and Canaries. The Iberian Water Frog and the Marsh Frog probably meet directly only as a result of introductions outside the natural range. The Iberian species is often greener, usually with an obvious yellow stripe down the centre of the back. It has some features in common with the Marsh Frog, such as its grey vocal sacs, small metatarsal tubercles, and whitish thighs.

**FOUND** *in unshaded wetlands, from still to flowing, lowland to montane, natural and artificial; is tolerant of polluted and brackish water.*

**NOTE**

*The Iberian Water Frog is one of the group of European frogs known as the water frogs, or green frogs. They are usually greener, noisier, and more aquatic, than the brown frogs; they also lack the dark eye-mask. But because of their odd breeding system, they are very confusing and identification needs great care.*

**SIZE** *Up to 8.5cm.*
**YOUNG** *Up to 10,000 eggs.*
**DIET** *Insects and other invertebrates; tadpoles and small fish.*
**STATUS** *Common.*
**SIMILAR SPECIES** *Edible Frog (p.214); Marsh Frogs (p.216), which rarely has a yellow stripe on its back; Graf's Hybrid Frog* (R. kl. grafi), *which has more fully webbed feet and larger metatarsal tubercles.*

# Marsh Frog

*Rana ridibunda* (Anura)

**FAVOURS** *waterbodies, such as ponds and lakes, with abundant vegetation.*

The largest native European frog, this species is also one of the noisiest, with loud croaks, rapidly repeated, and rising to a crescendo in chorus. Males have grey vocal sacs. Often seen basking on floating leaves, Marsh Frogs have whitish underparts, marbled in olive green. Highly aquatic, they take refuge in water when disturbed and hibernate underwater.

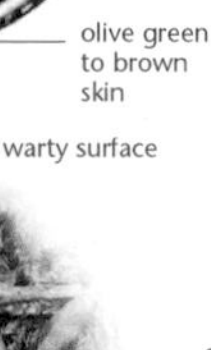

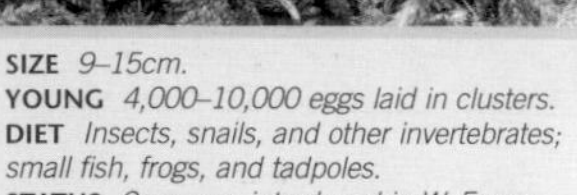

**SIZE** *9–15cm.*
**YOUNG** *4,000–10,000 eggs laid in clusters.*
**DIET** *Insects, snails, and other invertebrates; small fish, frogs, and tadpoles.*
**STATUS** *Common; introduced in W. Europe.*
**SIMILAR SPECIES** *Pool Frog and Edible Frog (p.214); Iberian Water Frog (p.215).*

small metatarsal tubercles

**HIND FOOT**

# Greek Marsh Frog

*Rana kurtmuelleri* (Anura)

**PREFERS** *large, open pools and lakes, tolerant of brackish and polluted water; where range overlaps with the Epirus Water Frog, often at higher altitudes.*

Found in mainland Greece and Albania, this frog is usually brown with a paler central stripe and irregular darker spots. However, range is usually the best clue to its identity. It often occurs (and hybridizes) with the Epirus Water Frog *(R. epeirotica)*; when disturbed, the Epirus Water Frog tends to flee on to land, but this species seeks refuge in water.

small metatarsal tubercle

**HIND FOOT**

**SIZE** *Up to 10cm.*
**YOUNG** *Around 5,000 eggs or more.*
**DIET** *Insects, snails, fish, and frogs.*
**STATUS** *Common.*
**SIMILAR SPECIES** *Karpathos* (R. cerigensis); *and Cretan* (R. cretensis) *Water Frogs; Albanian Pool Frog* (R. shqiperica).

# Glossary

Words in *italics* are defined elsewhere in the glossary.

**AESTIVATE** To undergo a period of dormancy during summer.

**ARBOREAL** Living fully or partly in trees.

**BALEEN** A fibrous substance that grows in the form of plates, which hang from a whale's upper jaw and are used to filter food from water.

**BLOWHOLE** The nostrils of whales, dolphins, and porpoises, positioned on top of their heads.

**BREACH** To leap out of water (usually the sea), landing back in it with a splash.

**CALCAR** In bats, a hollow *spur* extending from the hind limbs and helping to support the membrane between these.

**CREPUSCULAR** Active at twilight or before sunrise.

**CRITICALLY ENDANGERED** Facing an extremely high risk of global extinction.

**DECLINING** A reduction in the number of mature individuals of a species over a certain period.

**DIURNAL** Active during the day.

**DORSAL** On or near an animal's back.

**DREY** A squirrel's nest.

**ENDANGERED** Facing a very high risk of global extinction.

**GENUS** A unit of classification, grouping together closely related *species*, whose relationship is recognized by the same first name in the scientific terminology, e.g Triturus in Triturus alpestris.

**HYBRIDIZATION** The interbreeding of different *species* or *subspecies*.

**LANCET** Lance-shaped muscular projection on the face of a horseshoe-bat.

**LOCALLY COMMON** Abundant in particular parts of its range; even though it may be scarce or rare in other parts.

**LODGE** The den of certain animals, especially beavers.

**METATARSAL TUBERCLE** A prominent hard swelling under the hind foot of many species of frog and toad.

**MONTANE** Occurring in mountains, especially around and above the *tree-line*.

**NEAR-THREATENED** Likely to be at some risk of extinction if current trends continue.

**NOCTURNAL** Active at night.

**NOSE-LEAF** A thin, broad, membranous fold of skin on the nose of many species of horseshoe-bats. It varies greatly in size and form.

**NUTRIA** The pale brown fur of the coypu (the under fur).

**OCELLI** Eyelike spot with a light centre that contrasts with its surround.

**ORDER** A level used in classification. In the sequence of classification levels, an order forms part of a class, and is subdivided into one or more families.

**PAROTID GLAND** One of a pair of wart-like glands located behind the eyes in many amphibians, particularly conspicuous in toads. It may produce a noxious secretion.

**PREHENSILE** Able to curl around objects and grip them.

**RACE** See *subspecies*.

**RELICT SPECIES** A species that has survived while other related ones have become extinct.

**RIPARIAN** Occuring on the bank of a river.

**RUT** An annually recurring condition or period of sexual excitement and reproductive activity in male deer.

**SELLA** A saddle-shaped anatomical structure on the face of a horseshoe-bat.

**SETT** The burrow of a badger.

**SPECIES** A unit of classification that embraces a group of genetically similar individuals, members of which are capable of reproducing with one another and of producing viable offspring.

**SPUR** A pointed, projecting structure on the limbs.

**SUBSPECIES** A group of individuals that are distinctly different in appearance, and often geographically separated, from other members of the same species.

**TRAGUS** In most bats, a small cartilaginous flap in front of the external opening of the ear.

**TREE-LINE** The elevation in a mountainous region above which trees do not grow.

**TURGID** Swollen or distended.

**VULNERABLE** Facing a high risk of global extinction.

**WARREN** A series of interconnected underground tunnels in which rabbits live.

# Index

Scientific names often change as a result of reclassification – much of which is now based on genetic study. No attempt has been made here to include a full synonymy of scientific names. However, a number of the amphibians and reptiles have very recently been given new names. While the new names are used in the entries, the old names have been indexed here for reference.

## A



## B



## C

## N



## O



## P

# Acknowledgments

Dorling Kindersley would like to thank Neil Fletcher and Will Jones for their work as picture researchers and administrators and Miezan van Zyl for additional administrative assistance.

PICTURE CREDITS
Picture librarians: Claire Bowers and Lee Thompson.

Abbreviations key: a = above, b = bottom, c = centre, f =far, l = left, r = right, t = top.

The publishers would like to thank the following for their kind permission to reproduce the photographs.

**Yves Adams**: 037 cla; 040 cb; 061 bl; 062 cl; 127 clb; 205 cb; 207 cl. **Heather Angel/Natural Visions**: 98 ca. **Michel Aymerich**: 55 bc. **Lars Bergendorf**: 135 cb; 136 ca, cla, 137 cb; 138 car; 139 cfl; 144 ca; 147 ca, car; 206 car. **Dean Biggins, Denver Zoological Foundation**: 91bl. **David Bird**: 150 ca. **© Jonathan Bird/SeaPics.com**: 107 car. **Camille Braiden**: 189 tr. **Silvio Bruno**: 172 ca. **Peter Cairns**: 053 cr; 054 cb; 085 cbl; 095 ca; 097 c; 099 ca; 100 ca; 105 cra; 109 cb; 116 bl. **Mark Carwardine/Naturepl**: 107 br. **Cisca Castelijns/Foto Natura/FLPA**: 76 ca. **© Graeme Cresswell**: 106 cal. **George Dodge/Alamy**: 84 c. **Phil Dotson/Science Photo Library**: 103 cra. **Dudley Edmonson**: 063 cal; 087 cb; 088 car; 090 cra; 116 cra; 213 crb. **Gerry Ellis/Minden Pictures/FLPA**: 16 br. **Yossi Eshbol/FLPA**: 91 cra. **Neil Fletcher**: 015 tr; 016 tl; 017 tr; 018 tl, cfl; 019 tr; 021 tr; 022 tl; 023 tr; 024 tl; 026 tl; 027 ca; 029 tr; 030 tl; 031 tr; 032 tl; 033 tr; 034 tl, cfl; 035 tr; 037 cfr; 041 tr; 043 cfr; 044 cfl; 045 tr; 050 tl, cfl; 051 tr; 058 tl; 061 tr; 062 tl; 063 tr; 064 tl; 066 cfl; 072 tl; 087 cfl; 088 tl, bl; 090 tl; 092 cal; 102 tl; 103 tr; 104 tl; 105 cfr; 115 tr; 121 cbr; 128 crb; 129 tr; 132 tl; 133 tr; 134 tl; 137 tr, cfr; 141 tr; 143 cb; 147 tr; 152 tl; 158 tl; 161 cb; 200 bl; 204 cb; 209 c. **Paul Gale**: 108 ca. **Chris Gibson**: car. **Melvyn Grey**: 016 bcr; 046 cbr; 086 tl. **Morten Günther**: 155 cb. **© Clem Haagner/AfriPics.com**: 98 clb. **Daniel Heuclin/NHPA**: 26ca, bc. **Nigel Hicks**: 089 cbl; 117 cbl; 121 ca. **Josef Hlasek**: 015 bc; 053 cla; 056 car, cla; 057 cl, cr; 060 cal; 065 ca; 068 crb; 073 ca; 078 cra; 085 cla, cra; 095 bc; 096 cal; 100 cfl; 101 tr; 110 cb; 111 cb; 114 cra; 118 bl; 120 cla. **© Sascha Hooker/SeaPics.com**: 106bc. **Jon Hornbuckle**: 104 br; 108 bc, crb. **David Hosking/FLPA**: 17 cfr. **Geoffrey Kinns/Natural Visions**: 94 cra. **Barbara Klahr**: 147 bc. **Dr R. Kraft**: 19 cal, cb; 67car, bc. **Terry Longley**: 054 cla. **© Jose Luis Gomez de Francisco/Naturepl**: 50 bl; 71 ca. **Michal Maniakowski**: 56 bl. **Paolo Mazzei**: 123 cb; 124 cb; 125 cb; 129 cb; 130bc; 131 car; 132 cb; 133 bl; 136 cbr; 139 cra, ca, cb; 141 cbr; 142 cb; 143 cra, cla; 145 cb; 146 cbr, bl; 148 cal; 149 ca, cb, crb; 150 br; 151 cla; 152 cal; 153 clb; 154 crb; 155 cla, cra; 156 cbr; 158 cb; 159 cb; 160 cbl; 161 ca; 166 cra, cbl; 167 c, cb; 168 clb, crb; 170 clb; 171 cra, cfl, cb; 172 cbl; 174 cla, cr; 175 cb; 176 ca; 179 cr; 180 cal, cb; 183 cb, cfl, cla; 185 cb; 186 ca,cb; 187 ca; 190 ca; 191 bc; 192 ca, cb; 193 br; 194 cla, bc; 195 ca; 196 cl, cb; 197 tc, ca, cb, bl; 201 cla; 202 car; 204 ca, crb; 206 cla, cra, bl; 208 cla; 209 cbl; 213 clb; 216 cla. **Derek Middleton/FLPA**: 10 cla. **© Florian Möllers**: 93 cla. **MP O'Neill/Science Photo Library**: 128ca. **Dietmar Nill/Naturepl**: 189 cb. **Alan Outen**: 004 c; 016 cla; 049 cla; 082 ca; 113 cra; 120 cbr; 156 ca. **Oxford Scientific**: 25cra. **Doug Perrine/SeaPics.com**: 109 ca; 128 bl. **Iliara Pimpinelli**: 178 cb; 211 cra. **© Luis Quinta/SeaPics.com**: 103 bc. **Mike Read**: 050 tr, cra; 051 car, cb; 052 cal; 064 cal, cfr, cb; 073 cb; 080 cb; 081 cb; 086 ca; 088 cal; 092 cb; 096 cb; 101 ca; 104ca; 105 ca; 113 ca; 114 bl; 117 cfr; 118 cla; 119 car, bl; 120 cra; 121 cra. **Diego Reggianti**: 167 car; 169 ca, cb; 170 br; 179 cbl; 180 crb. **Robography/Alamy**: 111 ca. **Brian Rogers/Natural Visions**: 105 cbr. **Alexis Rosenfeld/Science Photo Library**: 105 bc. **© Mike Salisbury/SeaPics.com**: 102ca. **© Kevin Schafer/SeaPics.com**: 102cb. **Ty C. Smedes/smedesphoto.com**: 86 bc. **Roar Solheim**: 021 ca, cb; 060 cb; 066 cb; 094 cla. **Jeroen Speybroek**: 177 ca. **David Tipling/Alamy**: 118 ca. **Tom Uhlman/tomuphoto.com**: 62 bcl. **Jan Van Der Voort**: 002 c; 033 cfr; 056 cfl; 057 tr; 123 tr, ca; 124 ca, cra, crb; 126 cb; 129 car; 130 ca, cb; 131 cb; 132 ca; 133 ca, car, cb; 134 ca, cla, bc, crb; 135 cal; 136 br; 137 ca; 138 cb; 139 br; 140 ca, cb; 141 car, cfl; 142 cal,cl; 144 cla, cbl, bl; 145 cra, cl; 146 cla, cra; 148 cra, cbl, bl, br; 150 cb; 151 cra, cb, bl; 152 cra, clb, bcr; 153 ca, car, br; 154 cla, cra, bl; 157 ca, cb; 158 ca; 159 cal, cra; 160 ca, bl; 162 ca, cfr, cbl, crb; 163 c, car; 164 tl, cbl, ca; 165 cb, cfl, car; 166 cla, crb; 168 cra, cla; 170 cla, cra; 172 crb; 173 car, cl, cb; 175 car; 176 cla, clb, crb, bc; 177 cb, br; 178 ca, cla; 179 cra, cfl, br; 180 c, cbl; 184 ca, cb; 185 ca; 186 br; 187 clb; 188 ca, cb; 189 car; 190 cra, cbr, bl; 191 cal, cra; 193 tc, ca, clb; 195 cb, cbr; 200 cla, cra, bl; 201 cra, cb; 202 cla, cb; 203 cra, cla, bl, crb; 205 c; 206 br; 208 crb; 209 br; 210 cra, cla, clb, crb; 211cla, cb; 212 cla, cra, cbr, br; 213 cra, cla, cfl; 216 cra, cfr, crb, bl. **Rollin Verlinde**: 011 bl; 017 bc; 018 ca, cb; 020 cal, cb; 022c, cbl; 023 ca, br; 024 cla, cb; 027 cb; 029 cr; 030 cra, cla, bl, cbr; 031 ca, clb, crb; 032 cb; 033 ca, cb; 034 cla, bl; 035 ca, cbr; 036 cra, bc; 037 ca, br; 038 cfl; 039 cra, cbl; 040 car; 041 car, cb; 042 c, crb; 043 car, clb, br; 044 cal, bc; 045 car, cbl; 047 ca, cb; 048 ca, cb; 049 cb; 052 cbr; 055 cla; 058 ca; 059 ca, cb; 061 ca, cb, clb; 062 cal; 066 ca; 068 cla; 069 cra; 070 c, cb; 074 cl, crb; 075 cal, crb, clb; 076 cb; 077 ca, cb; 078 cb; 079 cb; 080 car; 081 ca; 082 cb; 083 clb; 087 cra; 090 bc; 094 bc; 100 br; 101 cb; 112 c; 113 bc; 115 c; 207 crb; 208 bl. **Visual&Written SL/Alamy**: 126 ca. **Dave Watts/NHPA**: 28 bc. **H. Willocx & R. Verlinde**: 071 cbl; 072 cbl. **Hugo Willocx**: 025 br; 028 car.